Succeed

Eureka Math®

Grade 5
Modules 3 & 4

Published by Great Minds®.

Printed in the U.S.A.
This book may be purchased from the publisher at eureka-math.org.
10 9 8 7 6 5 4 3

ISBN 978-1-64054-094-1

G5-M3-M4-S-06.2018

Learn • Practice • Succeed

Eureka Math® student materials for *A Story of Units*® (K–5) are available in the *Learn, Practice, Succeed* trio. This series supports differentiation and remediation while keeping student materials organized and accessible. Educators will find that the *Learn, Practice,* and *Succeed* series also offers coherent—and therefore, more effective—resources for Response to Intervention (RTI), extra practice, and summer learning.

Learn

Eureka Math Learn serves as a student's in-class companion where they show their thinking, share what they know, and watch their knowledge build every day. *Learn* assembles the daily classwork—Application Problems, Exit Tickets, Problem Sets, templates—in an easily stored and navigated volume.

Practice

Each *Eureka Math* lesson begins with a series of energetic, joyous fluency activities, including those found in *Eureka Math Practice.* Students who are fluent in their math facts can master more material more deeply. With *Practice,* students build competence in newly acquired skills and reinforce previous learning in preparation for the next lesson.

Together, *Learn* and *Practice* provide all the print materials students will use for their core math instruction.

Succeed

Eureka Math Succeed enables students to work individually toward mastery. These additional problem sets align lesson by lesson with classroom instruction, making them ideal for use as homework or extra practice. Each problem set is accompanied by a Homework Helper, a set of worked examples that illustrate how to solve similar problems.

Teachers and tutors can use *Succeed* books from prior grade levels as curriculum-consistent tools for filling gaps in foundational knowledge. Students will thrive and progress more quickly as familiar models facilitate connections to their current grade-level content.

Students, families, and educators:

Thank you for being part of the *Eureka Math*® community, where we celebrate the joy, wonder, and thrill of mathematics.

Nothing beats the satisfaction of success—the more competent students become, the greater their motivation and engagement. The *Eureka Math Succeed* book provides the guidance and extra practice students need to shore up foundational knowledge and build mastery with new material.

What is in the Succeed *book?*

Eureka Math Succeed books deliver supported practice sets that parallel the lessons of *A Story of Units*®. Each *Succeed* lesson begins with a set of worked examples, called *Homework Helpers*, that illustrate the modeling and reasoning the curriculum uses to build understanding. Next, students receive scaffolded practice through a series of problems carefully sequenced to begin from a place of confidence and add incremental complexity.

How should Succeed *be used?*

The collection of *Succeed* books can be used as differentiated instruction, practice, homework, or intervention. When coupled with *Affirm*®, *Eureka Math*'s digital assessment system, *Succeed* lessons enable educators to give targeted practice and to assess student progress. *Succeed*'s perfect alignment with the mathematical models and language used across *A Story of Units* ensures that students feel the connections and relevance to their daily instruction, whether they are working on foundational skills or getting extra practice on the current topic.

Where can I learn more about Eureka Math *resources?*

The Great Minds® team is committed to supporting students, families, and educators with an ever-growing library of resources, available at eureka-math.org. The website also offers inspiring stories of success in the *Eureka Math* community. Share your insights and accomplishments with fellow users by becoming a *Eureka Math* Champion.

Best wishes for a year filled with Eureka moments!

Jill Diniz

Jill Diniz
Director of Mathematics
Great Minds

Contents

Module 3: Addition and Subtraction of Fractions

Module 4: Multiplication and Division of Fractions and Decimal Fractions

Topic G: Division of Fractions and Decimal Fractions

Topic H: Interpretation of Numerical Expressions

Grade 5
Module 3

If I don't have the folded paper strip from class, I can cut a strip of paper about the length of this number line. I can fold it in 2 equal parts. Then, I can use it to label the number line.

1. Use the folded paper strip to mark points 0 and 1 above the number line and $\frac{0}{2}$, $\frac{1}{2}$, and $\frac{2}{2}$ below it.

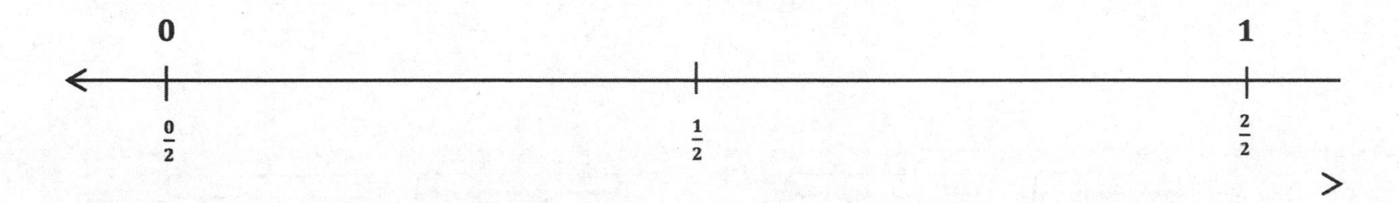

Draw one vertical line down the middle of each rectangle, creating two parts. Shade the left half of each. Partition with horizontal lines to show the equivalent fractions $\frac{2}{4}$, $\frac{3}{6}$, $\frac{4}{8}$, and $\frac{5}{10}$. Use multiplication to show the change in the units.

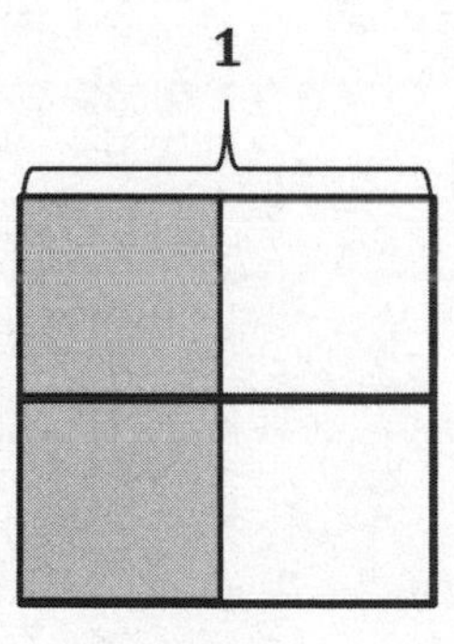

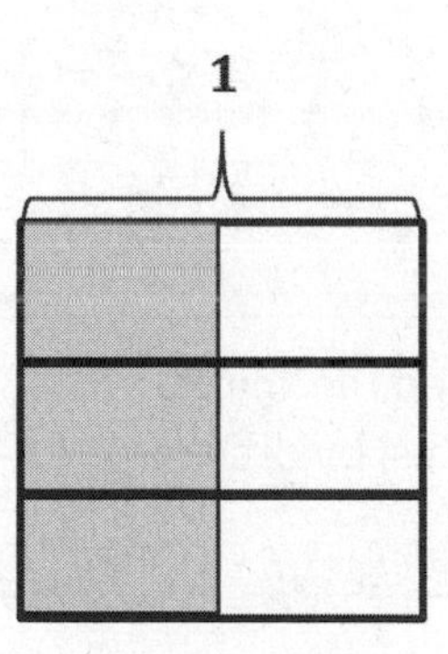

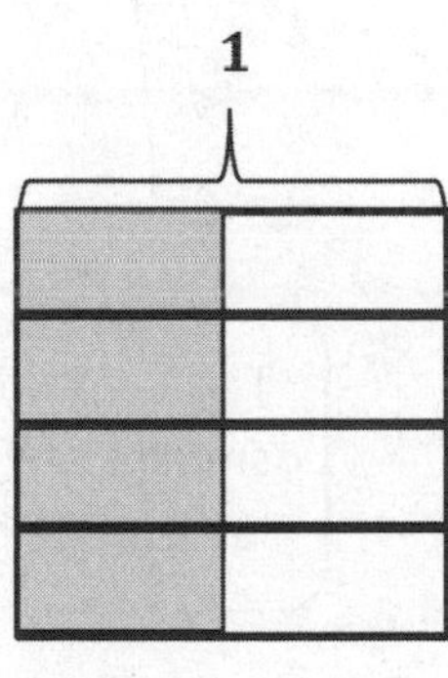

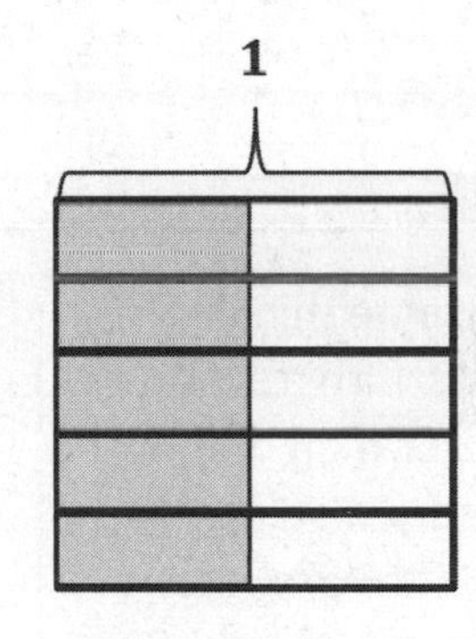

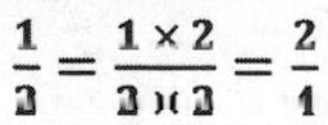

$\frac{1}{2} = \frac{1 \times 2}{2 \times 2} = \frac{2}{4}$

$\frac{1}{2} = \frac{1 \times 3}{2 \times 3} = \frac{3}{6}$

$\frac{1}{2} = \frac{1 \times 4}{2 \times 4} = \frac{4}{8}$

$\frac{1}{2} = \frac{1 \times 5}{2 \times 5} = \frac{5}{10}$

I started with one whole and divided it into halves by drawing 1 vertical line. I shaded 1 half. Then, I divided the halves into 2 equal parts by drawing a horizontal line. The shading shows me that $\frac{1}{2} = \frac{2}{4}$.

I did the same with the other models. I divided the halves into smaller units to make sixths, eighths, and tenths.

2. Continue the process, and model 2 equivalent fractions for 4 thirds. Estimate to mark the points on the number line.

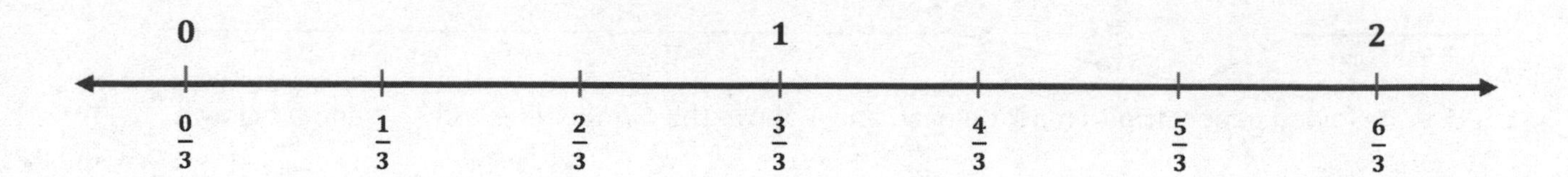

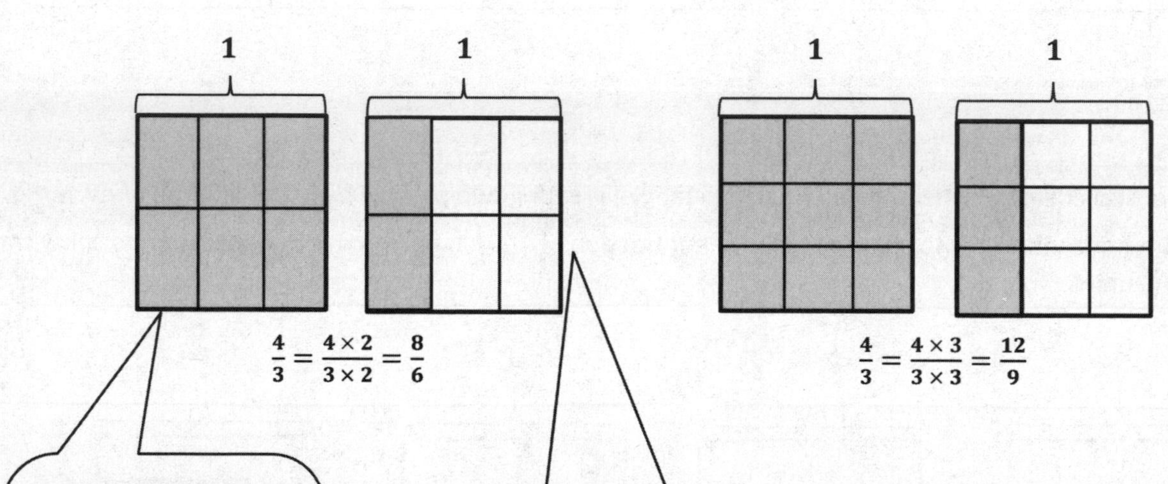

Name ______________________________ Date ______________

1. Use the folded paper strip to mark points 0 and 1 above the number line and $\frac{0}{3}$, $\frac{1}{3}$, $\frac{2}{3}$, and $\frac{3}{3}$ below it.

Draw two vertical lines to break each rectangle into thirds. Shade the left third of each. Partition with horizontal lines to show equivalent fractions. Use multiplication to show the change in the units.

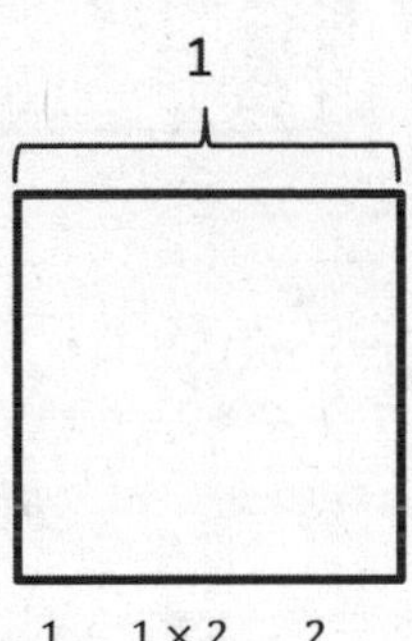
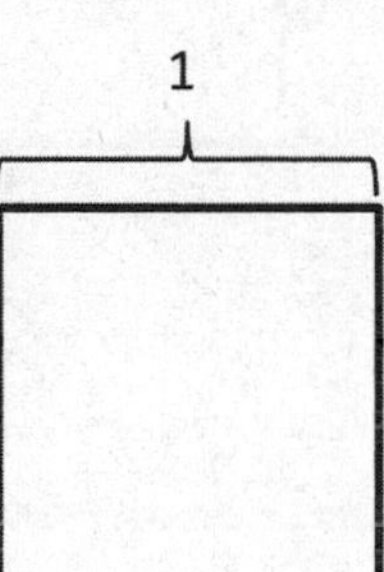
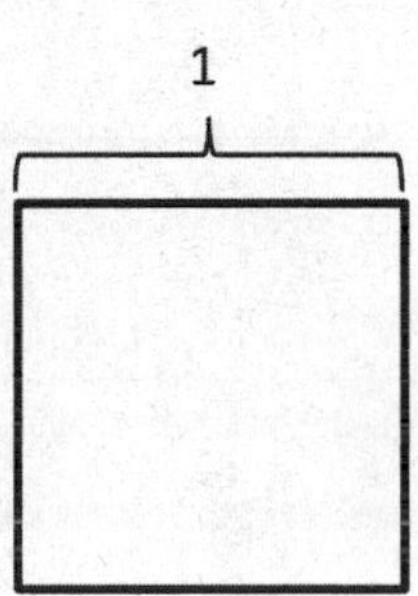
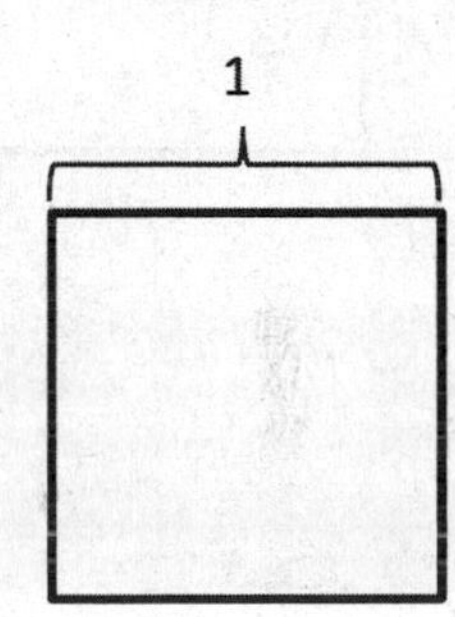

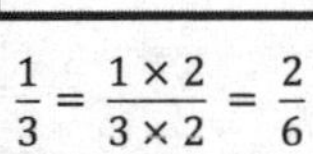
$\frac{1}{3} = \frac{1 \times 2}{3 \times 2} = \frac{2}{6}$

2. Use the folded paper strip to mark points 0 and 1 above the number line and $\frac{0}{4}$, $\frac{1}{4}$, $\frac{2}{4}$, $\frac{3}{4}$, and $\frac{4}{4}$ below it. Follow the same pattern as Problem 1 but with fourths.

1 1 1 1

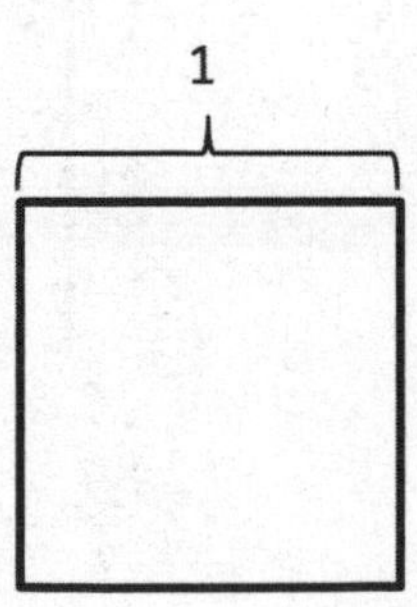
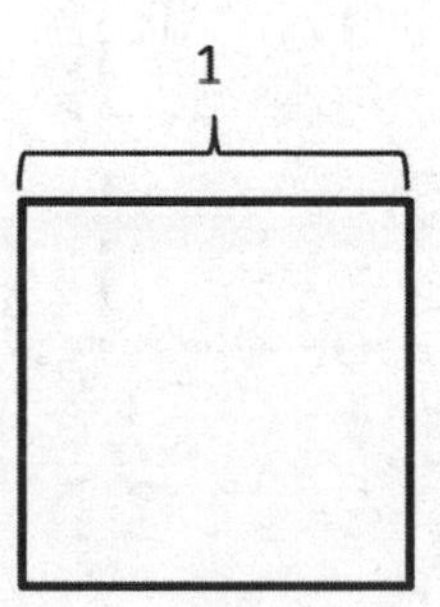
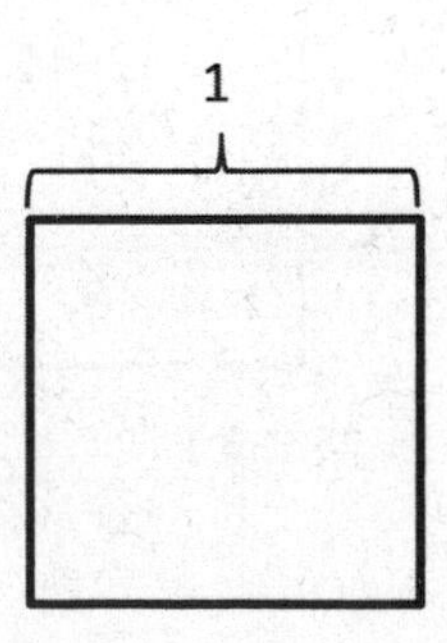

3. Continue the pattern with 4 fifths.

1 1 1 1

4. Continue the process, and model 2 equivalent fractions for 9 eighths. Estimate to mark the points on the number line.

1 1

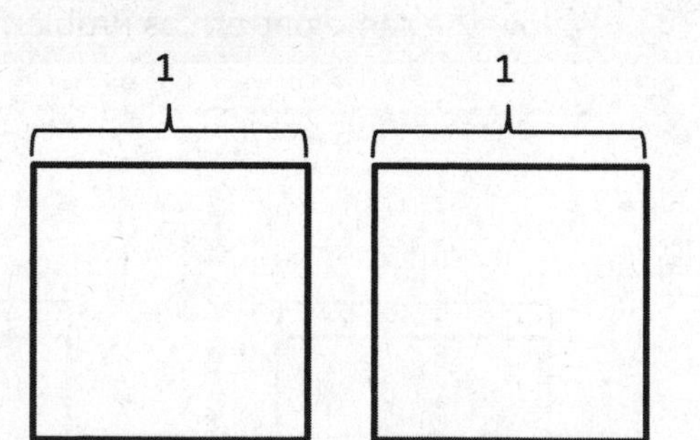

1. Show each expression on a number line. Solve.

   a. $\frac{1}{5}+\frac{1}{5}+\frac{2}{5}$

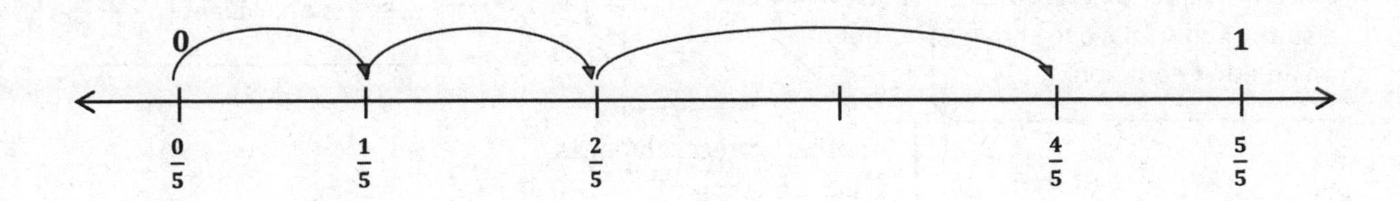

$$\frac{1}{5}+\frac{1}{5}+\frac{2}{5}=\frac{4}{5}$$

I'm not too concerned about making the jumps on the number line exactly proportional. The number line is just to help me visualize and calculate a solution.

   b. 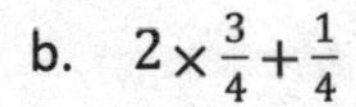 $2\times\frac{3}{4}+\frac{1}{4}$

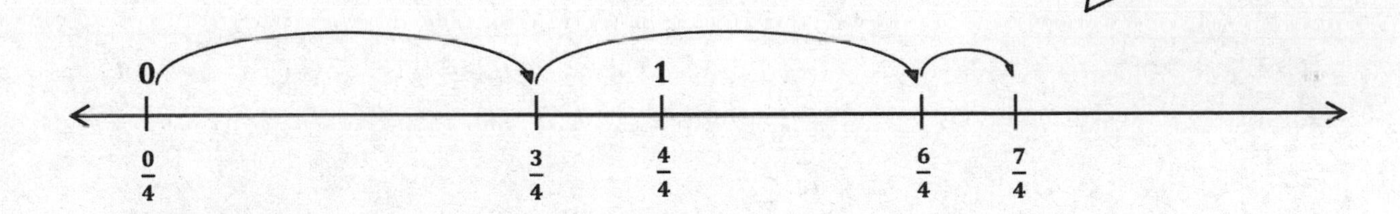

I can think of this problem in unit form: 2 times 3 fourths plus 1 fourth.

$$2\times\frac{3}{4}+\frac{1}{4}$$

$$=\frac{6}{4}+\frac{1}{4}=\frac{7}{4}$$

The answer doesn't have to be simplified. Writing either $\frac{7}{4}$ or $1\frac{3}{4}$ is correct.

2. Express $\frac{6}{5}$ as the sum of two or three equal fractional parts. Rewrite it as a multiplication equation, and then show it on a number line.

Since the directions asked for a sum, I know I have to show an addition equation.

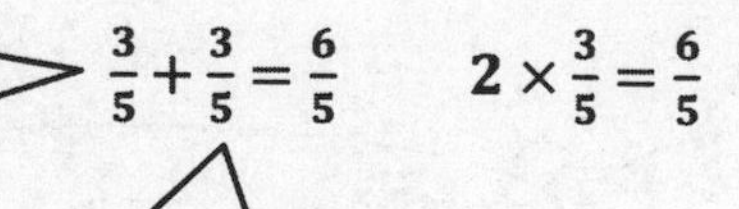

$2\times\frac{3}{5}$ is equivalent to $\frac{3}{5}+\frac{3}{5}$.

Another correct solution is $\frac{2}{5}+\frac{2}{5}+\frac{2}{5}=3\times\frac{2}{5}$.

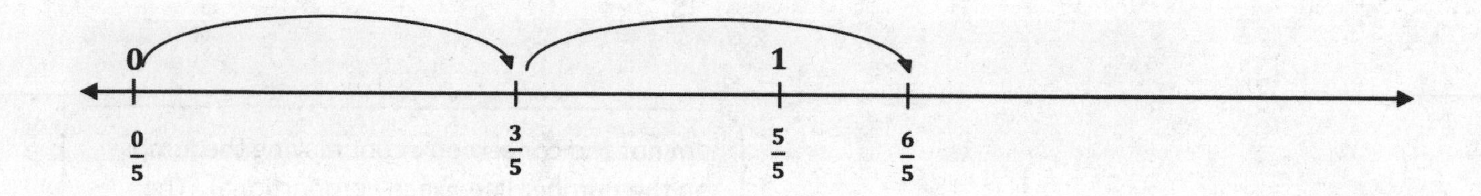

3. Express $\frac{7}{3}$ as the sum of a whole number and a fraction. Show on a number line.

$\frac{7}{3}=\frac{6}{3}+\frac{1}{3}$

$=2+\frac{1}{3}$

$=2\frac{1}{3}$

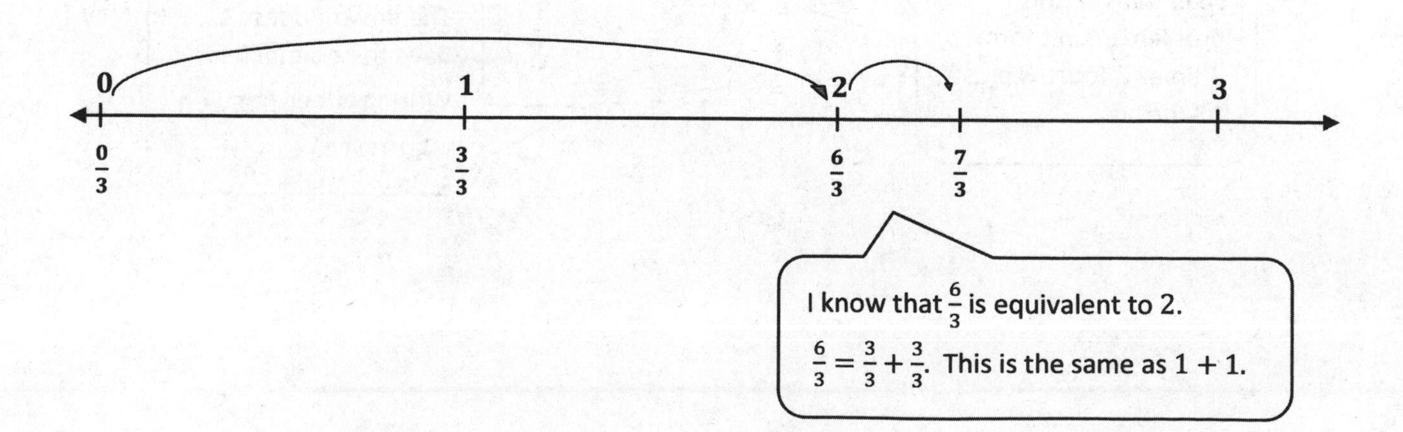

Name ________________________ Date ____________

1. Show each expression on a number line. Solve.

a. $\frac{4}{9} + \frac{1}{9}$

b. $\frac{1}{4} + \frac{1}{4} + \frac{1}{4} + \frac{1}{4}$

c. $\frac{2}{7} + \frac{2}{7} + \frac{2}{7}$

d. $2 \times \frac{3}{5} + \frac{1}{5}$

2. Express each fraction as the sum of two or three equal fractional parts. Rewrite each as a multiplication equation. Show Part (a) on a number line.

a. $\frac{6}{11}$

b. $\frac{9}{4}$

c. $\frac{12}{8}$

d. $\frac{27}{10}$

3. Express each of the following as the sum of a whole number and a fraction. Show Parts (c) and (d) on number lines.

a. $\frac{9}{5}$

b. $\frac{7}{2}$

c. $\frac{25}{7}$

d. $\frac{21}{9}$

4. Natalie sawed five boards of equal length to make a stool. Each was 9 tenths of a meter long. What is the total length of the boards she sawed? Express your answer as the sum of a whole number and the remaining fractional units. Draw a number line to represent the problem.

Draw a rectangular fraction model to find the sum. Simplify your answer, if possible.

a. $\frac{1}{2} + \frac{1}{3} = \frac{5}{6}$

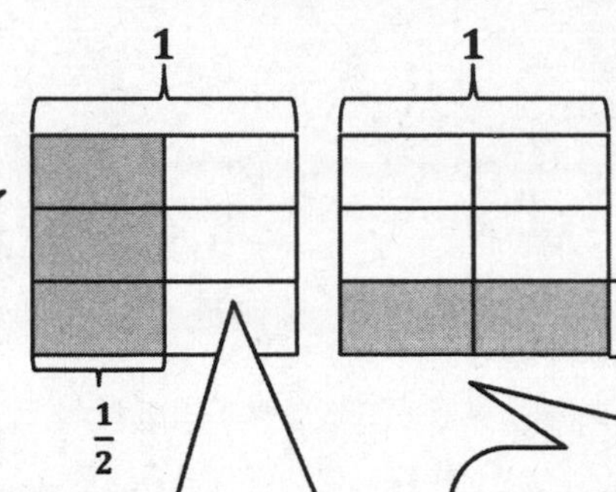

First, I make 2 identical wholes. I shade $\frac{1}{2}$ vertically. In the other whole I can show $\frac{1}{3}$ by drawing 2 horizontal lines.

I need to make like units in order to add. I partition the halves into sixths by drawing 2 horizontal lines.

$\frac{1}{2} = \frac{3}{6}$

I divide the thirds into sixths by drawing a vertical line. In both models, I have like units: sixths.

$\frac{1}{3} = \frac{2}{6}$

$$\frac{1}{2} + \frac{1}{3} = \frac{3}{6} + \frac{2}{6} = \frac{5}{6}$$

b. $\frac{2}{7} + \frac{2}{3} = \frac{20}{21}$

These addends are non-unit fractions because both have numerators greater than one.

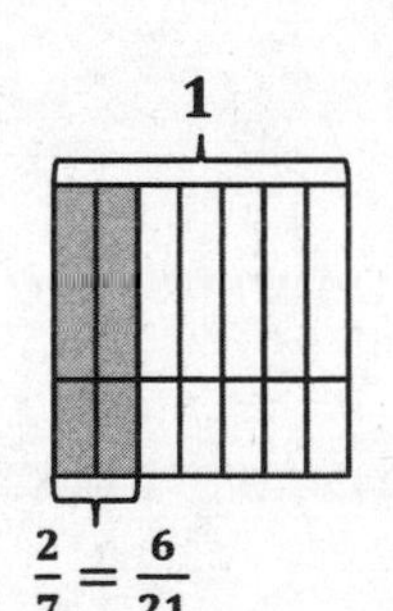

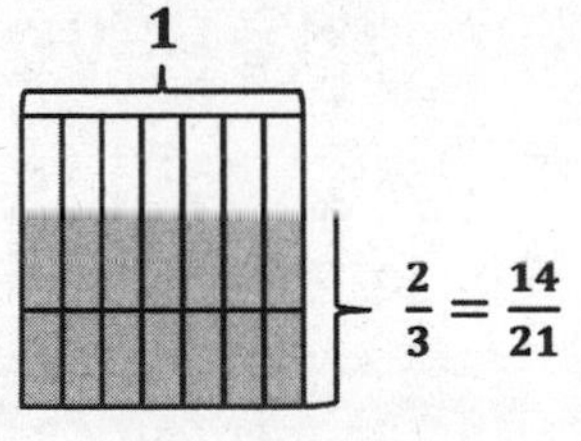

$$\frac{2}{7} + \frac{2}{3} = \frac{6}{21} + \frac{14}{21} = \frac{20}{21}$$

Name ______________________________ Date ________________

1. Draw a rectangular fraction model to find the sum. Simplify your answer, if possible.

a. $\frac{1}{4} + \frac{1}{3} =$

b. $\frac{1}{4} + \frac{1}{5} =$

c. $\frac{1}{4} + \frac{1}{6} =$

d. $\frac{1}{5} + \frac{1}{9} =$

e. $\frac{1}{4} + \frac{2}{5} =$

f. $\frac{3}{5} + \frac{3}{7} =$

Solve the following problems. Draw a picture, and write the number sentence that proves the answer. Simplify your answer, if possible.

2. Rajesh jogged $\frac{3}{4}$ mile and then walked $\frac{1}{6}$ mile to cool down. How far did he travel?

3. Cynthia completed $\frac{2}{3}$ of the items on her to-do list in the morning and finished $\frac{1}{8}$ of the items during her lunch break. What fraction of her to-do list is finished by the end of her lunch break? (Extension: What fraction of her to-do list does she still have to do after lunch?)

4. Sam read $\frac{2}{5}$ of her book over the weekend and $\frac{1}{6}$ of it on Monday. What fraction of the book has she read? What fraction of the book is left?

For the following problem, draw a picture using the rectangular fraction model, and write the answer. If possible, write your answer as a mixed number.

$\frac{1}{2}+\frac{3}{4}$

I need to make like units before adding.

My model shows me that $\frac{3}{4}=\frac{6}{8}$.

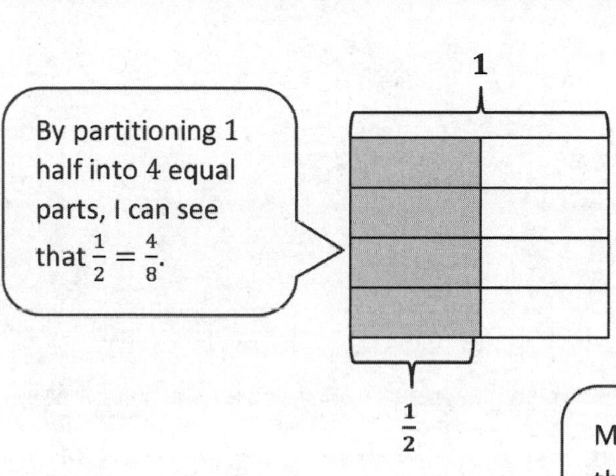

1

$\frac{3}{4}$

My solution of $1\frac{2}{8}$ makes sense. When I look at the fraction models and think about adding them together, I can see that they would make 1 whole and 2 eighths when combined.

$$\frac{1}{2}+\frac{3}{4}=\frac{4}{8}+\frac{6}{8}=\frac{10}{8}=1\frac{2}{8}$$

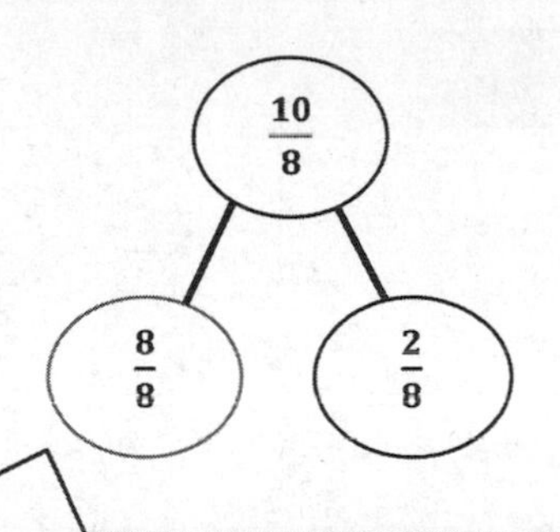

I don't need to express my solution in simplest form, but if wanted to, I could show that $1\frac{2}{8}=1\frac{1}{4}$.

I can use a number bond to rename $\frac{10}{8}$ as a mixed number. This part-part-whole model shows that 10 eighths is composed of 8 eighths and 2 eighths.

Name ______________________ Date ____________

1. For the following problems, draw a picture using the rectangular fraction model and write the answer. When possible, write your answer as a mixed number.

a. $\frac{3}{4} + \frac{1}{3} =$

b. $\frac{3}{4} + \frac{2}{3} =$

c. $\frac{1}{3} + \frac{3}{5} =$

d. $\frac{5}{6} + \frac{1}{2} =$

e. $\frac{2}{3} + \frac{5}{6} =$

f. $\frac{4}{3} + \frac{4}{7} =$

Solve the following problems. Draw a picture, and write the number sentence that proves the answer. Simplify your answer, if possible.

2. Sam made $\frac{2}{3}$ liter of punch and $\frac{3}{4}$ liter of tea to take to a party. How many liters of beverages did Sam bring to the party?

3. Mr. Sinofsky used $\frac{5}{8}$ of a tank of gas on a trip to visit relatives for the weekend and another 1 half of a tank commuting to work the next week. He then took another weekend trip and used $\frac{1}{4}$ tank of gas. How many tanks of gas did Mr. Sinofsky use altogether?

1. Find the difference. Use a rectangular fraction model to find a common unit. Simplify your answer, if possible.

$\frac{2}{3} - \frac{1}{4} = \frac{5}{12}$

In order to subtract fourths from thirds, I need to find like units.

I draw 2 vertical lines to partition my model into thirds and shade 2 of them to show the fraction $\frac{2}{3}$.

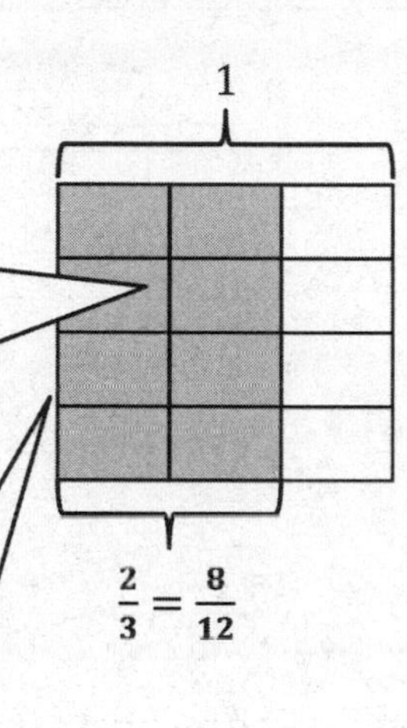

1

I draw 3 horizontal lines to partition my model into fourths and shade 1 of them to show the fraction $\frac{1}{4}$.

$\frac{1}{4} = \frac{3}{12}$

In order to make like units, or common denominators, I draw 3 horizontal lines to partition the model into 12 equal parts. Now, I can see that $\frac{2}{3} = \frac{8}{12}$.

I still can't subtract. Fourths and twelfths are different units. But, I can draw 2 vertical lines to partition the model into 12 equal parts. Now, I have equal units and can see that $\frac{1}{4} = \frac{3}{12}$.

$$\frac{2}{3} - \frac{1}{4} = \frac{8}{12} - \frac{3}{12} = \frac{5}{12}$$

Once I have like units, the subtraction is simple. I know that 8 minus 3 is equal to 5, so I can think of this in unit form very simply.

8 twelfths − 3 twelfths = 5 twelfths

2. Lisbeth needs $\frac{1}{3}$ of a tablespoon of spice for a baking recipe. She has $\frac{5}{6}$ of a tablespoon in her pantry. How much spice will Lisbeth have after baking?

I'll need to subtract $\frac{1}{3}$ from $\frac{5}{6}$ to find out how much remains.

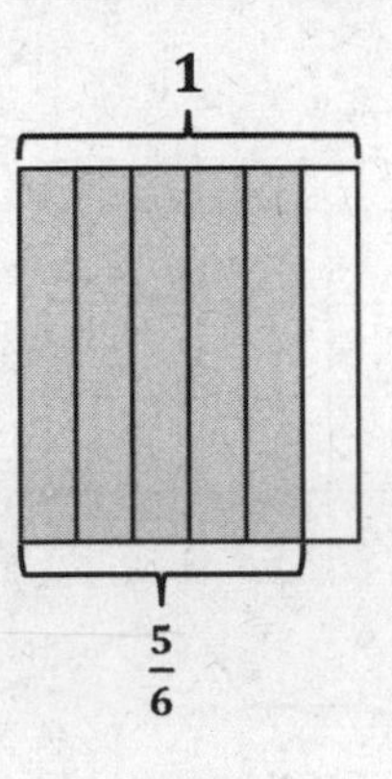

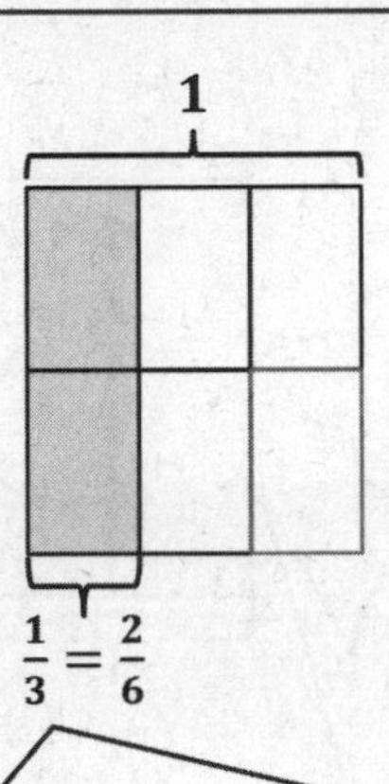

This was interesting! After drawing the $\frac{5}{6}$ that Lisbeth has in her pantry, I realized that thirds and sixths are related units. In this problem, I could leave $\frac{5}{6}$ as is and only rename the thirds as sixths to find a common unit.

$\frac{5}{6} - \frac{1}{3} = \frac{5}{6} - \frac{2}{6} = \frac{3}{6}$

Lisbeth will have $\frac{3}{6}$ of a tablespoon of spice after baking.

I could also express $\frac{3}{6}$ as $\frac{1}{2}$ because they are equivalent fractions, but I don't have to.

In order to finish the problem, I must make a statement to answer the question.

Name ______________________________ Date ______________

1. The picture below shows $\frac{3}{4}$ of the rectangle shaded. Use the picture to show how to create an equivalent fraction for $\frac{3}{4}$, and then subtract $\frac{1}{3}$.

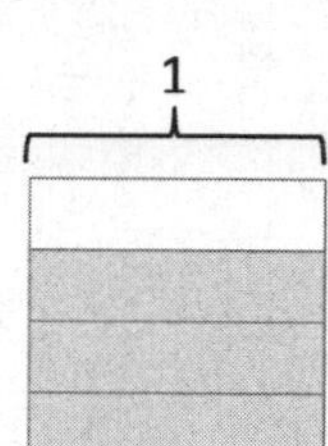

$\frac{3}{4} - \frac{1}{3} =$

2. Find the difference. Use a rectangular fraction model to find common denominators. Simplify your answer, if possible.

a. $\frac{5}{6} - \frac{1}{3} =$

b. $\frac{2}{3} - \frac{1}{2} =$

c. $\frac{5}{6} - \frac{1}{4} =$

d. $\frac{4}{5} - \frac{1}{2} =$

e. $\frac{2}{3} - \frac{2}{5} =$

f. $\frac{5}{7} - \frac{2}{3} =$

3. Robin used $\frac{1}{4}$ of a pound of butter to make a cake. Before she started, she had $\frac{7}{8}$ of a pound of butter. How much butter did Robin have when she was done baking? Give your answer as a fraction of a pound.

4. Katrina needs $\frac{3}{5}$ kilogram of flour for a recipe. Her mother has $\frac{3}{7}$ kilogram of flour in her pantry. Is this enough flour for the recipe? If not, how much more will she need?

For the following problems, draw a picture using the rectangular fraction model, and write the answer. Simplify your answer, if possible.

a. $\frac{4}{3} - \frac{1}{2} = \frac{5}{6}$

In order to subtract halves from thirds, I'll need to find a common unit. I can rename them both as a number of sixths.

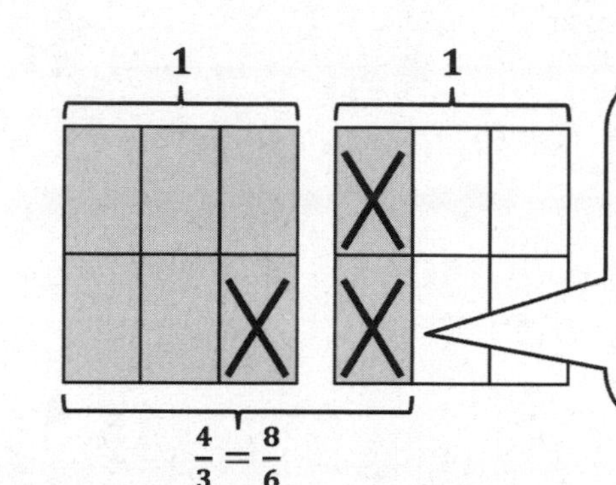

I can cross out the $\frac{3}{6}$ that I'm subtracting to see the $\frac{5}{6}$ that represents the difference.

$\frac{4}{3} - \frac{1}{2} = \frac{8}{6} - \frac{3}{6} = \frac{5}{6}$

$\frac{4}{3} = \frac{3}{3} + \frac{1}{3} = 1 + \frac{1}{3}$ and $\frac{8}{6} = \frac{6}{6} + \frac{2}{6} = 1 + \frac{2}{6}$

b. $1\frac{2}{3} - \frac{3}{4} = \frac{11}{12}$

In order to subtract fourths from thirds, I'll need to find a common unit. I can rename them both as a number of twelfths.

This time, I'll subtract $\frac{3}{4}$ (or $\frac{9}{12}$) all at once from the 1 (or the $\frac{12}{12}$).

Then, in order to find the difference, I can add these $\frac{3}{12}$ to the $\frac{8}{12}$ in the fraction model to the right.

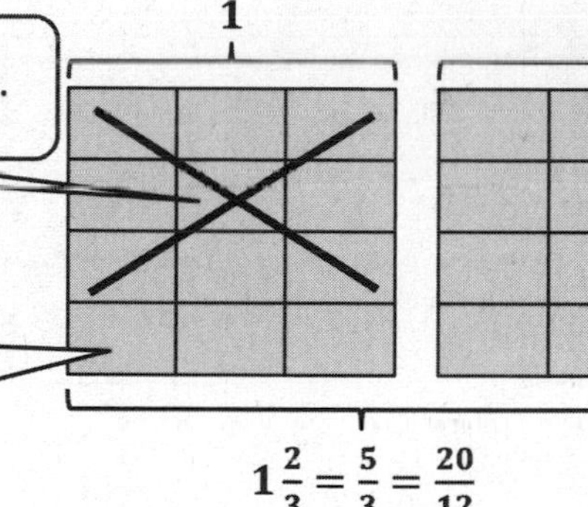

I can use the fraction model and this number bond to help me see that $1\frac{2}{3}$ is composed of $\frac{12}{12}$ and $\frac{8}{12}$.

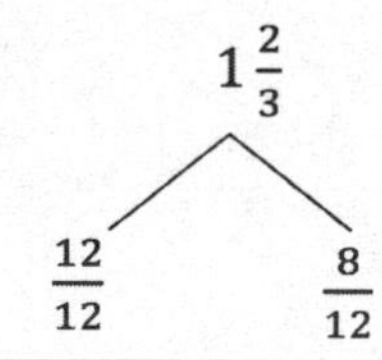

$1\frac{2}{3} - \frac{3}{4} = \frac{3}{12} + \frac{8}{12} = \frac{11}{12}$

Name ______________________________ Date ______________

1. For the following problems, draw a picture using the rectangular fraction model and write the answer. Simplify your answer, if possible.

a. $1 - \frac{5}{6} =$

b. $\frac{3}{2} - \frac{5}{6} =$

c. $\frac{4}{3} - \frac{5}{7} =$

d. $1\frac{1}{8} - \frac{3}{5} =$

e. $1\frac{2}{5} - \frac{3}{4} =$

f. $1\frac{5}{6} - \frac{7}{8} =$

g. $\frac{9}{7} - \frac{3}{4} =$

h. $1\frac{3}{12} - \frac{2}{3} =$

2. Sam had $1\frac{1}{2}$ m of rope. He cut off $\frac{5}{8}$ m and used it for a project. How much rope does Sam have left?

3. Jackson had $1\frac{3}{8}$ kg of fertilizer. He used some to fertilize a flower bed, and he only had $\frac{2}{3}$ kg left. How much fertilizer was used in the flower bed?

RDW means "Read, Draw, Write." I **read** the problem several times. I **draw** something each time I read. I remember to **write** the answer to the question.

Solve the word problems using the RDW strategy.

1. Rosie has a collection of comic books. She gave $\frac{1}{2}$ of them to her brother. Rosie gave $\frac{1}{6}$ of them to her friend, and she kept the rest. How much of the collection did Rosie keep for herself?

If I subtract $\frac{1}{2}$ and $\frac{1}{6}$ from 1, I can find how much of the collection Rosie kept for herself.

I can draw a tape diagram to model this problem.

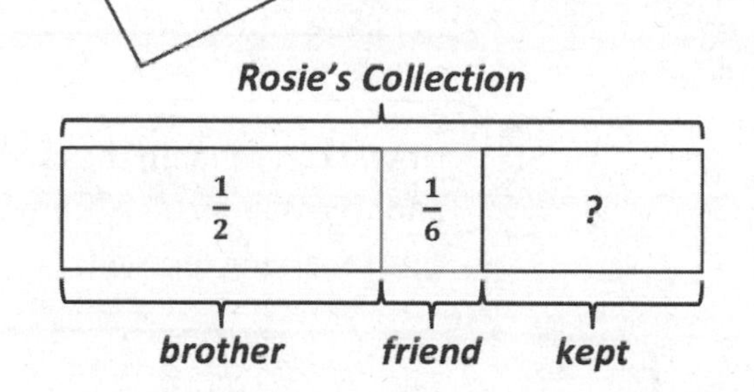

$1 - \frac{1}{2} - \frac{1}{6}$

$= \frac{1}{2} - \frac{1}{6}$

$= \frac{3}{6} - \frac{1}{6}$

$= \frac{2}{6}$

I've been doing so much of this that now I can rename some fractions in my head. I know that $\frac{1}{2} = \frac{3}{6}$.

Rosie kept $\frac{2}{6}$ or $\frac{1}{3}$ of the collection for herself.

When I think of this another way, I know that my solution makes sense. I can think $\frac{1}{2} + \frac{1}{6} +$ "how much more" is equal to 1?

$$\frac{1}{2} + \frac{1}{6} + ? = 1 \rightarrow \frac{3}{6} + \frac{1}{6} + \frac{2}{6} = \frac{6}{6} = 1$$

2. Ken ran for $\frac{1}{4}$ mile. Peggy ran $\frac{1}{3}$ mile farther than Ken. How far did they run altogether?

Ken $\frac{1}{4}$ mi

Peggy $\frac{1}{3}$ mi

?

To find the distance they ran altogether, I'll add Ken's distance ($\frac{1}{4}$ mile) to Peggy's distance ($\frac{1}{4}$ mile + $\frac{1}{3}$ mile).

My tape diagram shows that Peggy ran the same distance as Ken plus $\frac{1}{3}$ mile farther.

$\frac{1}{4}+\frac{1}{4}+\frac{1}{3}$

$=\frac{1}{2}+\frac{1}{3}$

$=\frac{3}{6}+\frac{2}{6}$

$=\frac{5}{6}$

I could rename all of these as a number of twelfths, but I know that $\frac{1}{4}+\frac{1}{4}=\frac{2}{4}$, which is equal to $\frac{1}{2}$.

Now, I can rename these halves and thirds as sixths. I can do this renaming mentally!

Ken and Peggy ran $\frac{5}{6}$ mile altogether.

EUREKA MATH

Name ______________________________ Date ________________

Solve the word problems using the RDW strategy. Show all of your work.

1. Christine baked a pumpkin pie. She ate $\frac{1}{6}$ of the pie. Her brother ate $\frac{1}{3}$ of it and gave the leftovers to his friends. What fraction of the pie did he give to his friends?

2. Liang went to the bookstore. He spent $\frac{1}{3}$ of his money on a pen and $\frac{4}{7}$ of it on books. What fraction of his money did he have left?

3. Tiffany bought $\frac{2}{5}$ kg of cherries. Linda bought $\frac{1}{10}$ kg of cherries less than Tiffany. How many kilograms of cherries did they buy altogether?

4. Mr. Rivas bought a can of paint. He used $\frac{3}{8}$ of it to paint a bookshelf. He used $\frac{1}{4}$ of it to paint a wagon. He used some of it to paint a birdhouse and has $\frac{1}{8}$ of the paint left. How much paint did he use for the birdhouse?

5. Ribbon A is $\frac{1}{3}$ m long. It is $\frac{2}{5}$ m shorter than Ribbon B. What's the total length of the two ribbons?

1. Add or subtract. Draw a number line to model your solution.

   a. $9\frac{1}{3} + 6 = \mathbf{15\frac{1}{3}}$

   $9\frac{1}{3}$ is the same as $9 + \frac{1}{3}$. I can add the whole numbers, $9 + 6 = 15$, and then add the fraction, $15 + \frac{1}{3} = 15\frac{1}{3}$.

   I can model this addition using a number line. I'll start at 0 and add 9.

   I add 6 to get to 15.

   Then, I add $\frac{1}{3}$ to get to $15\frac{1}{3}$.

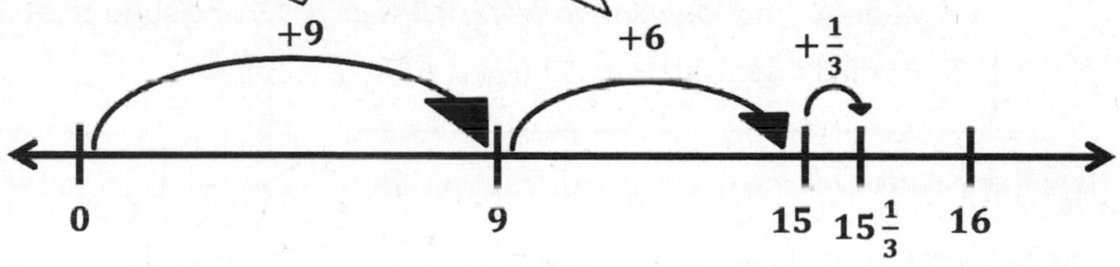

   $+9$ $+6$ $+\frac{1}{3}$

   0 9 15 $15\frac{1}{3}$ 16

   b. $18 - 13\frac{3}{4} = \mathbf{4\frac{1}{4}}$

   $13\frac{3}{4}$ is the same as $13 + \frac{3}{4}$. I can subtract the whole numbers first, $18 - 13 = 5$. Then, I can subtract the fraction, $5 - \frac{3}{4} = 4\frac{1}{4}$.

   I start at 18 and subtract 13 to get 5. Then, I subtract $\frac{3}{4}$ to get $4\frac{1}{4}$.

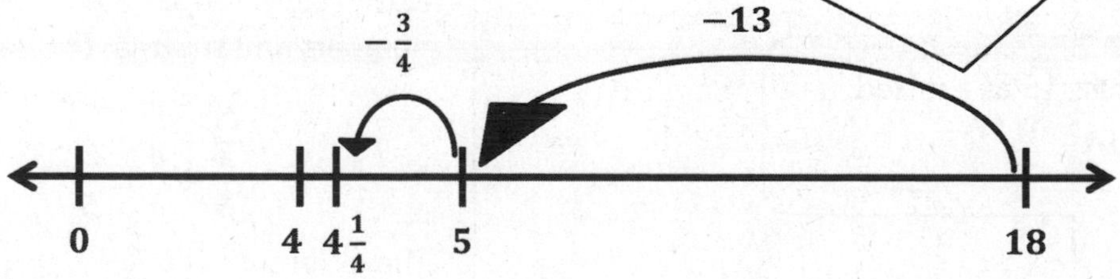

   $-\frac{3}{4}$ -13

   0 4 $4\frac{1}{4}$ 5 18

2. The total length of two strings is 15 meters. If one string is $8\frac{3}{5}$ meters long, what is the length of the other string?

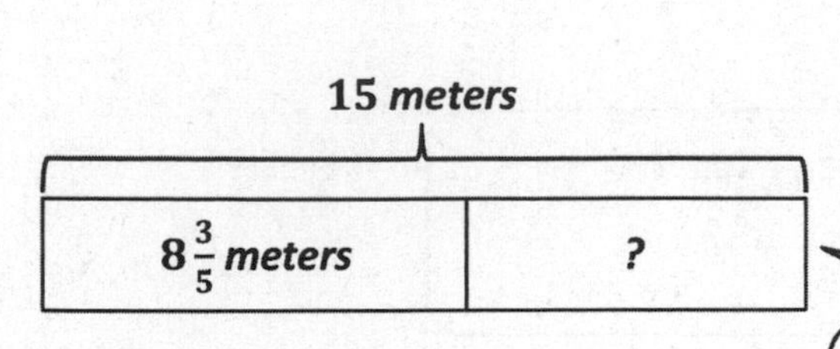

I can use subtraction, $15 - 8\frac{3}{5}$, to find the length of the other string.

My tape diagram models this word problem. I need to find the length of the missing part.

$$15 - 8\frac{3}{5} = 6\frac{2}{5}$$

I can draw a number line to solve. I'll start at 15 and subtract 8 to get 7. Then, I'll subtract $\frac{3}{5}$ to get $6\frac{2}{5}$.

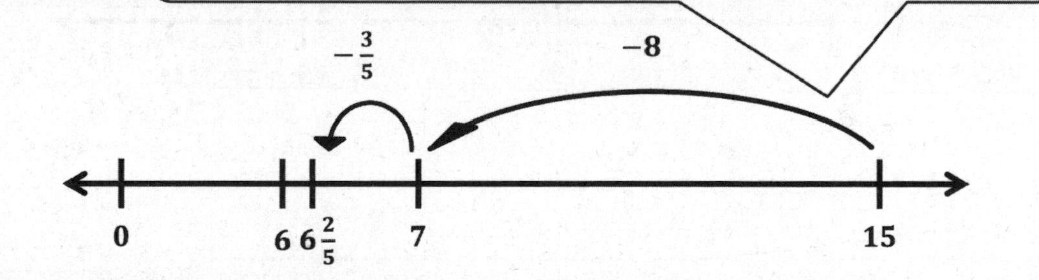

The length of the other string is $6\frac{2}{5}$ meters.

Below is an alternative method to solve this problem.

I can express 15 as a mixed number, $14\frac{5}{5}$.

Now, I can subtract the whole numbers and subtract the fractions.

$$14 - 8 = 6$$

$$\frac{5}{5} - \frac{3}{5} = \frac{2}{5}$$

The difference is $6\frac{2}{5}$.

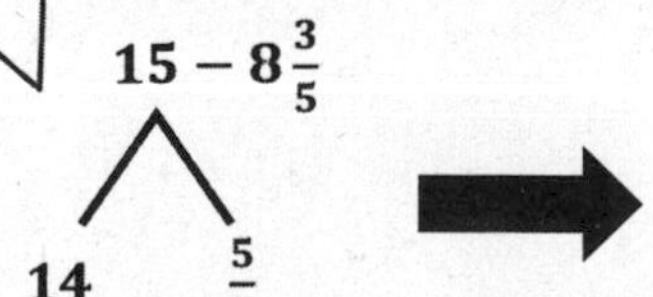

$$14\frac{5}{5} - 8\frac{3}{5} = 6\frac{2}{5}$$

Name ______________________________ Date ________________

1. Add or subtract.

a. $3 + 1\frac{1}{4} =$

b. $2 - 1\frac{5}{8} =$

c. $5\frac{2}{5} + 2\frac{3}{5} =$

d. $4 - 2\frac{5}{7} =$

e. $8\frac{4}{5} + 7 =$

f. $18 - 15\frac{3}{4} =$

g. $16 + 18\frac{5}{6} =$

h. $100 - 50\frac{3}{8} =$

2. The total length of two ribbons is 13 meters. If one ribbon is $7\frac{5}{8}$ meters long, what is the length of the other ribbon?

3. It took Sandy two hours to jog 13 miles. She ran $7\frac{1}{2}$ miles in the first hour. How far did she run during the second hour?

4. Andre says that $5\frac{3}{4} + 2\frac{1}{4} = 7\frac{1}{2}$ because $7\frac{4}{8} = 7\frac{1}{2}$. Identify his mistake. Draw a picture to prove that he is wrong.

1. First, make like units, and then add.

The denominators here are thirds and fifths. I can skip count to find a like unit.

3: $3, 6, 9, 12, \mathbf{15}, 18, \ldots$
5: $5, 10, \mathbf{15}, 20, \ldots$

15 is a multiple of both 3 and 5, so I can make like units of fifteenths.

I can multiply both the numerator and the denominator by 5 to rename $\frac{1}{3}$ as a number of fifteenths.

$\frac{1 \times 5}{3 \times 5} = \frac{5}{15}$

$$\frac{1}{3} + \frac{2}{5} = \left(\frac{1 \times 5}{3 \times 5}\right) + \left(\frac{2 \times 3}{5 \times 3}\right)$$
$$= \frac{5}{15} + \frac{6}{15}$$
$$= \frac{11}{15}$$

I can multiply both the numerator and the denominator by 3 to rename $\frac{2}{5}$ as a number of fifteenths.

$\frac{2 \times 3}{5 \times 3} = \frac{6}{15}$

5 fifteenths + 6 fifteenths = 11 fifteenths

The denominators here are sixths and eighths. I can skip count to find a like unit.
6: 6, 12, 18, **24**, 30, ...
8: 8, 16, **24**, 32, ...
24 is a multiple of both 6 and 8, so I can make like units of twenty-fourths.

I can multiply both the numerator and the denominator by 4 to rename $\frac{5}{6}$ as a number of twenty-fourths.
$\frac{5 \times 4}{6 \times 4} = \frac{20}{24}$

b. $\frac{5}{6} + \frac{3}{8} = \left(\frac{5 \times 4}{6 \times 4}\right) + \left(\frac{3 \times 3}{8 \times 3}\right)$

$= \frac{20}{24} + \frac{9}{24}$

$= \frac{29}{24}$

$= \frac{24}{24} + \frac{5}{24}$

$= 1\frac{5}{24}$

I can multiply both the numerator and the denominator by 3 to rename $\frac{3}{8}$ as a number of twenty-fourths.
$\frac{3 \times 3}{8 \times 3} = \frac{9}{24}$

$\frac{29}{24}$ is the same as $\frac{24}{24}$ plus $\frac{5}{24}$, or $1\frac{5}{24}$.

The like unit for ninths and halves is eighteenths.

c. $\frac{4}{9} + 1\frac{1}{2} = \left(\frac{4 \times 2}{9 \times 2}\right) + \left(\frac{1 \times 9}{2 \times 9}\right) + 1$

$= \frac{8}{18} + \frac{9}{18} + 1$

$= \frac{17}{18} + 1$

$= 1\frac{17}{18}$

I can add the 1 after adding the fractions.

$\frac{17}{18}$ plus 1 is the same as the mixed number $1\frac{17}{18}$.

2. On Tuesday, Karol spent $\frac{3}{4}$ of one hour on reading homework and $\frac{1}{3}$ of one hour on math homework. How much time did Karol spend doing her reading and math homework on Tuesday?

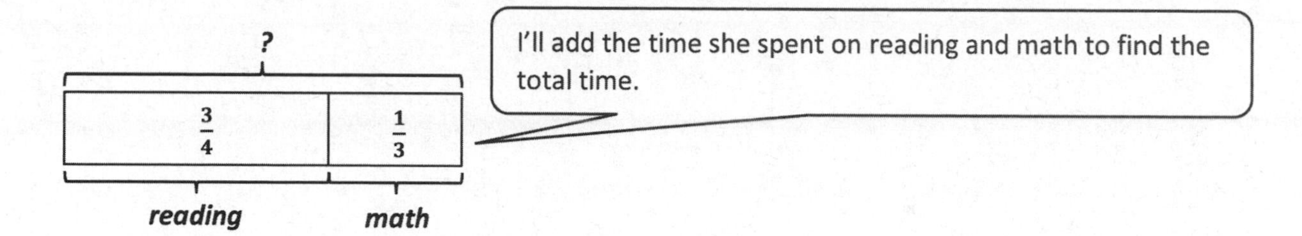

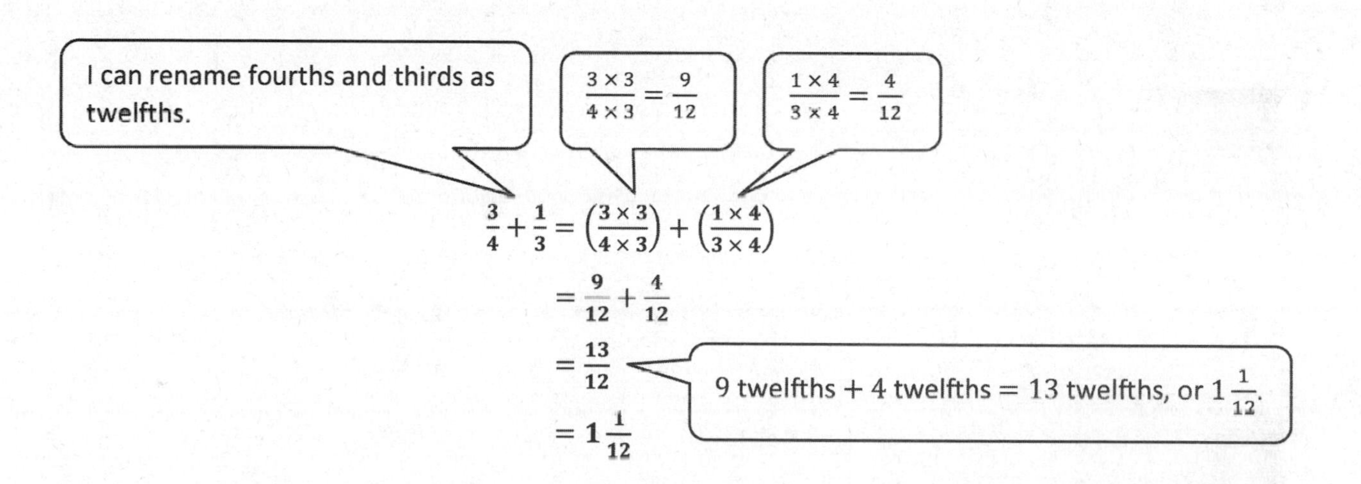

$$\frac{3}{4} + \frac{1}{3} = \left(\frac{3 \times 3}{4 \times 3}\right) + \left(\frac{1 \times 4}{3 \times 4}\right)$$
$$= \frac{9}{12} + \frac{4}{12}$$
$$= \frac{13}{12}$$
$$= 1\frac{1}{12}$$

Karol spent $1\frac{1}{12}$ hours doing her reading and math homework.

Name ______________________________ Date ______________

1. Make like units, and then add.

a. $\frac{3}{5} + \frac{1}{3} =$

b. $\frac{3}{5} + \frac{1}{11} =$

c. $\frac{2}{9} + \frac{5}{6} =$

d. $\frac{2}{5} + \frac{1}{4} + \frac{1}{10} =$

e. $\frac{1}{3} + \frac{7}{5} =$

f. $\frac{5}{8} + \frac{7}{12} =$

g. $1\frac{1}{3}+\frac{3}{4}=$

h. $\frac{5}{6}+1\frac{1}{4}=$

2. On Monday, Ka practiced guitar for $\frac{2}{3}$ of one hour. When she finished, she practiced piano for $\frac{3}{4}$ of one hour. How much time did Ka spend practicing instruments on Monday?

3. Ms. How bought a bag of rice for dinner. She used $\frac{3}{5}$ kg of the rice and still had $2\frac{1}{4}$ kg left. How heavy was the bag of rice that Ms. How bought?

4. Joe spends $\frac{2}{5}$ of his money on a jacket and $\frac{3}{8}$ of his money on a shirt. He spends the rest on a pair of pants. What fraction of his money does he use to buy the pants?

I'll add the whole numbers first and then add the fractions. $4 + 2 = 6$

1. Add.

a. $4\frac{2}{5} + 2\frac{1}{3} = \mathbf{6 + \frac{2}{5} + \frac{1}{3}}$

$= \mathbf{6 + \left(\frac{2 \times 3}{5 \times 3}\right) + \left(\frac{1 \times 5}{3 \times 5}\right)}$

I need to make like units before adding.

$= \mathbf{6 + \frac{6}{15} + \frac{5}{15}}$

$= \mathbf{6 + \frac{11}{15}}$

$= \mathbf{6\frac{11}{15}}$

I can rename these fractions as a number of fifteenths. $\frac{2}{5} = \frac{6}{15}$, and $\frac{1}{3} = \frac{5}{15}$.

The sum is $6\frac{11}{15}$.

I'll add the whole numbers together. $5 + 10 = 15$.

b. $5\frac{2}{7} + 10\frac{3}{4} = \mathbf{15 + \frac{2}{7} + \frac{3}{4}}$

$= \mathbf{15 + \left(\frac{2 \times 4}{7 \times 4}\right) + \left(\frac{3 \times 7}{4 \times 7}\right)}$

When I look at $\frac{2}{7}$ and $\frac{3}{4}$, I decide to use 28 as the common unit, which will be the new denominator.
$\frac{2}{7} = \frac{8}{28}$
$\frac{3}{4} = \frac{21}{28}$

$= \mathbf{15 + \frac{8}{28} + \frac{21}{28}}$

$= \mathbf{15 + \frac{29}{28}}$

$= \mathbf{15 + \frac{28}{28} + \frac{1}{28}}$

I know $\frac{29}{28}$ is more than 1. So, I'll rewrite $\frac{29}{28}$ as $\frac{28}{28} + \frac{1}{28}$.

$= \mathbf{16\frac{1}{28}}$

The sum is $16\frac{1}{28}$.

2. Jillian bought some ribbon. She used $3\frac{3}{4}$ meters for an art project and had $5\frac{1}{10}$ meters left. What was the original length of the ribbon?

I can add to find the original length of the ribbon.

I draw a tape diagram and label the used ribbon $3\frac{3}{4}$ meters and the leftover ribbon $5\frac{1}{10}$ meters.

I label the whole ribbon with a question mark because that's what I'm trying to find.

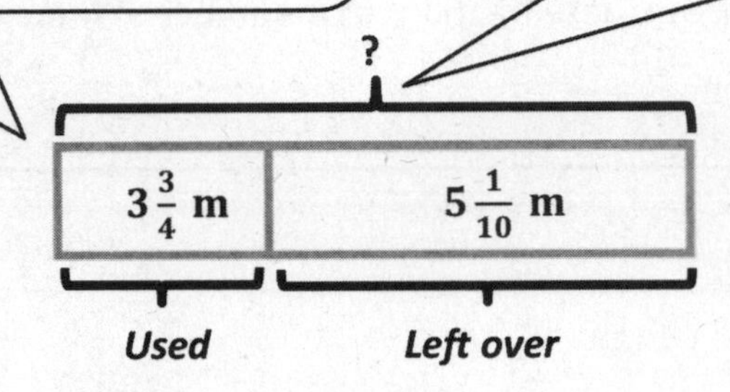

I'll add 3 plus 5 to get 8.

I need to rename fourths and tenths as a common unit before adding. When I skip-count, I know that 20 is a multiple of both 4 and 10.

$$3\frac{3}{4} + 5\frac{1}{10} = 8 + \frac{3}{4} + \frac{1}{10}$$
$$= 8 + \left(\frac{3 \times 5}{4 \times 5}\right) + \left(\frac{1 \times 2}{10 \times 2}\right)$$
$$= 8 + \frac{15}{20} + \frac{2}{20}$$
$$= 8\frac{17}{20}$$

$\frac{3}{4} = \frac{15}{20}$, and $\frac{1}{10} = \frac{2}{20}$.

The original length of the ribbon was $8\frac{17}{20}$ meters.

Name ______________________ Date ____________

1. Add.

a. $2\frac{1}{2} + 1\frac{1}{5} =$

b. $2\frac{1}{2} + 1\frac{3}{5} =$

c. $1\frac{1}{5} + 3\frac{1}{3} =$

d. $3\frac{2}{3} + 1\frac{3}{5} =$

e. $2\frac{1}{3} + 4\frac{4}{7} =$

f. $3\frac{5}{7} + 4\frac{2}{3} =$

g. $15\frac{1}{5} + 4\frac{3}{8} =$

h. $18\frac{3}{8} + 2\frac{2}{5} =$

2. Angela practiced piano for $2\frac{1}{2}$ hours on Friday, $2\frac{1}{3}$ hours on Saturday, and $3\frac{2}{3}$ hours on Sunday. How much time did Angela practice piano during the weekend?

3. String A is $3\frac{5}{6}$ meters long. String B is $2\frac{1}{4}$ meters long. What's the total length of both strings?

4. Matt says that $5 - 1\frac{1}{4}$ will be more than 4, since 5 – 1 is 4. Draw a picture to prove that Matt is wrong.

1. Generate equivalent fractions to get like units and then, subtract.

   a. $\frac{3}{4} - \frac{1}{3}$

      $= \frac{9}{12} - \frac{4}{12}$

      $= \frac{5}{12}$

      I can rename fourths and thirds as twelfths in order to subtract. $\frac{3}{4} = \frac{9}{12}$ and $\frac{1}{3} = \frac{4}{12}$.

      9 twelfths − 4 twelfths = 5 twelfths

   b. $3\frac{4}{5} - 2\frac{1}{2}$

      I can rename halves and fifths as tenths to subtract. I can solve this problem in several different ways.

      Method 1:

      $3\frac{4}{5} - 2\frac{1}{2}$

      $= 3\frac{8}{10} - 2\frac{5}{10}$

      $= 1\frac{3}{10}$

      I can rewrite the mixed numbers with a common denominator of 10.
      $3\frac{4}{5} = 3\frac{8}{10}$, and $2\frac{1}{2} = 2\frac{5}{10}$.

      Now, I can subtract the whole numbers and then the fractions.
      $3 - 2 = 1$, and $\frac{8}{10} - \frac{5}{10} = \frac{3}{10}$.

      The answer is $1 + \frac{3}{10}$, or $1\frac{3}{10}$.

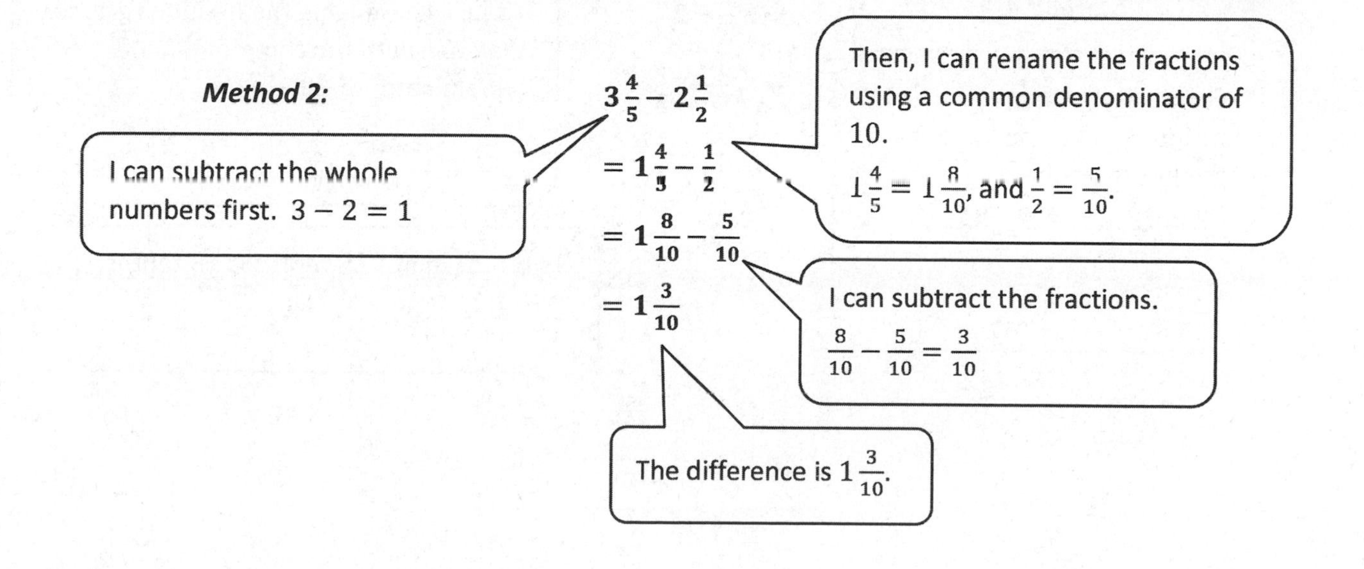

Method 3:

I can also decompose $3\frac{4}{5}$ into two parts using a number bond.

$$3\frac{4}{5} - 2\frac{1}{2}$$

(number bond: 3 and $\frac{4}{5}$)

Now, I can easily subtract $2\frac{1}{2}$ from 3.
$3 - 2\frac{1}{2} = \frac{1}{2}$

After subtracting $2\frac{1}{2}$, I can add the remaining fractions, $\frac{1}{2}$ and $\frac{4}{5}$.

$$= \frac{1}{2} + \frac{4}{5}$$

$$= \frac{5}{10} + \frac{8}{10}$$

I can rename these fractions as tenths in order to add.
$\frac{1}{2} = \frac{5}{10}$, and $\frac{4}{5} = \frac{8}{10}$.

$$= \frac{13}{10}$$

$$= 1\frac{3}{10}$$

The sum of 5 tenths and 8 tenths is 13 tenths. $\frac{13}{10} = \frac{10}{10} + \frac{3}{10} = 1\frac{3}{10}$

Method 4:

$$3\frac{4}{5} - 2\frac{1}{2}$$

I could also rename the mixed numbers as fractions greater than one.
$3\frac{4}{5} = \frac{15}{5} + \frac{4}{5} = \frac{19}{5}$, and
$2\frac{1}{2} = \frac{4}{2} + \frac{1}{2} = \frac{5}{2}$.

$$= \frac{19}{5} - \frac{5}{2}$$

$$= \frac{38}{10} - \frac{25}{10}$$

Then, I can rename the fractions greater than one with the common denominator of 10.
$\frac{19}{5} = \frac{38}{10}$, and $\frac{5}{2} = \frac{25}{10}$.

$$= \frac{13}{10}$$

$$= 1\frac{3}{10}$$

38 tenths minus 25 tenths is 13 tenths.
$\frac{13}{10} = \frac{10}{10} + \frac{3}{10} = 1\frac{3}{10}$.

Name ____________________________________ Date ____________________

1. Generate equivalent fractions to get like units. Then, subtract.

a. $\frac{1}{2} - \frac{1}{5} =$

b. $\frac{7}{8} - \frac{1}{3} =$

c. $\frac{7}{10} - \frac{3}{5} =$

d. $1\frac{5}{6} - \frac{2}{3} =$

e. $2\frac{1}{4} - 1\frac{1}{5} =$

f. $5\frac{6}{7} - 3\frac{2}{3} =$

g. $15\frac{7}{8} - 5\frac{3}{4} =$

h. $15\frac{5}{8} - 3\frac{1}{3} =$

2. Sandy ate $\frac{1}{6}$ of a candy bar. John ate $\frac{3}{4}$ of it. How much more of the candy bar did John eat than Sandy?

3. $4\frac{1}{2}$ yards of cloth are needed to make a woman's dress. $2\frac{2}{7}$ yards of cloth are needed to make a girl's dress. How much more cloth is needed to make a woman's dress than a girl's dress?

4. Bill reads $\frac{1}{5}$ of a book on Monday. He reads $\frac{2}{3}$ of the book on Tuesday. If he finishes reading the book on Wednesday, what fraction of the book did he read on Wednesday?

5. Tank A has a capacity of 9.5 gallons. $6\frac{1}{3}$ gallons of the tank's water are poured out. How many gallons of water are left in the tank?

1. Subtract.

I can subtract these mixed numbers using a variety of strategies.

a. $3\frac{1}{4} - 2\frac{1}{3}$

I can rename these fractions as twelfths in order to subtract.

Method 1:

$$3\frac{1}{4} - 2\frac{1}{3}$$
$$= 1\frac{1}{4} - \frac{1}{3}$$
$$= 1\frac{3}{12} - \frac{4}{12}$$
$$= \frac{15}{12} - \frac{4}{12}$$
$$= \frac{11}{12}$$

I can subtract the whole numbers. $3 - 2 = 1$

I can rename the fractions with a common unit of 12. $1\frac{1}{4} = 1\frac{3}{12}$, and $\frac{1}{3} = \frac{4}{12}$.

I can't subtract the fraction $\frac{4}{12}$ from $\frac{3}{12}$, so I can rename $1\frac{3}{12}$ as a fraction greater than one, $\frac{15}{12}$.

15 twelfths − 4 twelfths = 11 twelfths

Method 2:

$$3\frac{1}{4} - 2\frac{1}{3}$$

Number bond: $3\frac{1}{4}$ → 3 and $\frac{1}{4}$

$$\frac{2}{3} + \frac{1}{4}$$
$$= \frac{8}{12} + \frac{3}{12}$$
$$= \frac{11}{12}$$

Or, I could decompose $3\frac{1}{4}$ into two parts with a number bond.

Now, I can easily subtract $2\frac{1}{3}$ from 3. $3 - 2\frac{1}{3} = \frac{2}{3}$

After subtracting $2\frac{1}{3}$, I can add the remaining fractions, $\frac{2}{3}$ and $\frac{1}{4}$.

I can rename these fractions as twelfths in order to add. $\frac{2}{3} = \frac{8}{12}$, and $\frac{1}{4} = \frac{3}{12}$.

The sum of 8 twelfths and 3 twelfths is 11 twelfths.

Or, I could rename both mixed numbers as fractions greater than one.

$3\frac{1}{4}=\frac{13}{4}$, and $2\frac{1}{3}=\frac{7}{3}$.

Method 3:

$$\mathbf{3\frac{1}{4}-2\frac{1}{3}}$$
$$=\mathbf{\frac{13}{4}-\frac{7}{3}}$$
$$=\mathbf{\frac{39}{12}-\frac{28}{12}}$$
$$=\mathbf{\frac{11}{12}}$$

And, I can rename the fractions greater than one using the common unit twelfths.

$\frac{13}{4}=\frac{39}{12}$, and $\frac{7}{3}=\frac{28}{12}$.

39 twelfths minus 28 twelfths is equal to 11 twelfths.

b. $19\frac{1}{3}-4\frac{6}{7}$

Method 1:

$$\mathbf{19\frac{1}{3}-4\frac{6}{7}}$$
$$=\mathbf{15\frac{1}{3}-\frac{6}{7}}$$
$$=\mathbf{15\frac{7}{21}-\frac{18}{21}}$$
$$=\mathbf{14\frac{28}{21}-\frac{18}{21}}$$
$$=\mathbf{14\frac{10}{21}}$$

I need to make a common unit before subtracting. I can rename these fractions using a denominator of 21.

I can subtract the whole numbers, $19-4=15$

$15\frac{7}{21}=14+1+\frac{7}{21}$

$=14+\frac{21}{21}+\frac{7}{21}$

$=14+\frac{28}{21}$

$=14\frac{28}{21}$

I can't subtract $\frac{18}{21}$ from $\frac{7}{21}$, so I rename $15\frac{7}{21}$ as $14\frac{28}{21}$.

Method 2:

I want to subtract $4\frac{6}{7}$ from 5, so I can decompose $19\frac{1}{3}$ into two parts with this number bond.

$5-4\frac{6}{7}=\frac{1}{7}$

Now, I need to combine $\frac{1}{7}$ with the remaining part, $14\frac{1}{3}$.

$$\mathbf{19\frac{1}{3}-4\frac{6}{7}}=\mathbf{\frac{1}{7}+14\frac{1}{3}}$$
$$=\mathbf{\frac{3}{21}+14\frac{7}{21}}$$
$$=\mathbf{14\frac{10}{21}}$$

Number bond: $\mathbf{19\frac{1}{3}}$ → $\mathbf{14\frac{1}{3}}$ and $\mathbf{5}$

In order to add, I'll rename these fractions using a common denominator of 21.

Name ________________________ Date ____________

1. Subtract.

a. $3\frac{1}{4} - 2\frac{1}{3} =$

b. $3\frac{2}{3} - 2\frac{3}{4} =$

c. $6\frac{1}{5} - 4\frac{1}{4} =$

d. $6\frac{3}{5} - 4\frac{3}{4} =$

e. $5\frac{2}{7} - 4\frac{1}{3} =$

f. $8\frac{2}{3} - 3\frac{5}{7} =$

g. $18\frac{3}{4} - 5\frac{7}{8} =$

h. $17\frac{1}{5} - 2\frac{5}{8} =$

2. Tony wrote the following:

$$7\frac{1}{4} - 3\frac{3}{4} = 4\frac{1}{4} - \frac{3}{4}.$$

Is Tony's statement correct? Draw a number line to support your answer.

3. Ms. Sanger blended $8\frac{3}{4}$ gallons of iced tea with some lemonade for a picnic. If there were $13\frac{2}{5}$ gallons of the beverage, how many gallons of lemonade did she use?

4. A carpenter has $10\frac{1}{2}$ feet of wooden plank. He cuts off $4\frac{1}{4}$ feet to replace the slat of a deck and $3\frac{2}{3}$ feet to repair a bannister. He uses the rest of the plank to fix a stair. How many feet of wood does the carpenter use to fix the stair?

1. Are the following expressions greater than or less than 1? Circle the correct answer.

   a. $\frac{1}{2}+\frac{3}{5}$ (greater than 1) less than 1

   I know that $\frac{1}{2}$ plus $\frac{1}{2}$ is exactly 1. I also know that $\frac{3}{5}$ is greater than $\frac{1}{2}$. Therefore, $\frac{1}{2}$ plus a number greater than $\frac{1}{2}$ must be greater than 1.

   b. $3\frac{1}{4}-2\frac{2}{3}$ greater than 1 (less than 1)

   I know that $3-2=1$, so this expression is the same as $1\frac{1}{4}-\frac{2}{3}$. I also know that $\frac{2}{3}$ is greater than $\frac{1}{4}$. Therefore, if I were to subtract $\frac{2}{3}$ from $1\frac{1}{4}$, the difference would be less than 1.

2. Are the following expressions greater than or less than $\frac{1}{2}$? Circle the correct answer.

   $\frac{1}{3}+\frac{1}{4}$ (greater than $\frac{1}{2}$) less than $\frac{1}{2}$

   I know that $\frac{1}{4}$ plus $\frac{1}{4}$ is exactly $\frac{1}{2}$. I also know that $\frac{1}{3}$ is greater than $\frac{1}{4}$. Therefore, $\frac{1}{4}$ plus a number greater than $\frac{1}{4}$ must be greater than $\frac{1}{2}$.

3. Use $>$, $<$, or $=$ to make the following statement true.

   $6\frac{3}{4}$ ___>___ $2\frac{4}{5}+3\frac{1}{3}$

   I know that 3 plus $3\frac{1}{3}$ is equal to $6\frac{1}{3}$, which is less than $6\frac{3}{4}$.

   Therefore, a number less than 3 plus $3\frac{1}{3}$ is definitely going to be less than $6\frac{3}{4}$.

Name ______________________________ Date ______________

1. Are the following expressions greater than or less than 1? Circle the correct answer.

 a. $\frac{1}{2}+\frac{4}{9}$ greater than 1 less than 1

 b. $\frac{5}{8}+\frac{3}{5}$ greater than 1 less than 1

 c. $1\frac{1}{5}-\frac{1}{3}$ greater than 1 less than 1

 d. $4\frac{3}{5}-3\frac{3}{4}$ greater than 1 less than 1

2. Are the following expressions greater than or less than $\frac{1}{2}$? Circle the correct answer.

 a. $\frac{1}{5}+\frac{1}{4}$ greater than $\frac{1}{2}$ less than $\frac{1}{2}$

 b. $\frac{6}{7}-\frac{1}{6}$ greater than $\frac{1}{2}$ less than $\frac{1}{2}$

 c. $1\frac{1}{7}-\frac{5}{6}$ greater than $\frac{1}{2}$ less than $\frac{1}{2}$

 d. $\frac{4}{7}+\frac{1}{8}$ greater than $\frac{1}{2}$ less than $\frac{1}{2}$

3. Use >, <, or = to make the following statements true.

 a. $5\frac{4}{5}+2\frac{2}{3}$ ______ $8\frac{3}{4}$

 b. $3\frac{4}{7}-2\frac{3}{5}$ ______ $1\frac{4}{7}+\frac{3}{5}$

 c. $4\frac{1}{2}+1\frac{4}{9}$ ______ $5+\frac{13}{18}$

 d. $10\frac{3}{8}-7\frac{3}{5}$ ______ $3\frac{3}{8}+\frac{3}{5}$

4. Is it true that $5\frac{2}{3} - 3\frac{3}{4} = 1 + \frac{2}{3} + \frac{3}{4}$? Prove your answer.

5. A tree limb hangs $5\frac{1}{4}$ feet from a telephone wire. The city trims back the branch *before* it grows within $2\frac{1}{2}$ feet of the wire. Will the city allow the tree to grow $2\frac{3}{4}$ more feet?

6. Mr. Kreider wants to paint two doors and several shutters. It takes $2\frac{1}{8}$ gallons of paint to coat each door and $1\frac{3}{5}$ gallons of paint to coat all of his shutters. If Mr. Kreider buys three 2-gallon cans of paint, does he have enough to complete the job?

1. Rearrange the terms so that you can add or subtract mentally, and then solve.

   a. $2\frac{1}{3} - \frac{3}{5} + \frac{2}{3} = \left(\mathbf{2\frac{1}{3} + \frac{2}{3}}\right) - \mathbf{\frac{3}{5}}$

   $= \mathbf{3 - \frac{3}{5}}$

   $= \mathbf{2\frac{2}{5}}$

   The associative property allows me to rearrange these terms so that I can add the like units first.

   Wow! This is actually a really basic problem now!

   b. $8\frac{3}{4} - 2\frac{2}{5} - 1\frac{1}{5} - \frac{3}{4} = \left(\mathbf{8\frac{3}{4} - \frac{3}{4}}\right) - \left(\mathbf{2\frac{2}{5} + 1\frac{1}{5}}\right)$

   This expression has fourths and fifths. I can use the associative property to rearrange the like units together.

   $= \mathbf{8 - 3\frac{3}{5}}$

   $= \mathbf{5 - \frac{3}{5}}$

   $= \mathbf{4\frac{2}{5}}$

   Subtracting $2\frac{2}{5}$ and then subtracting $1\frac{1}{5}$ is the same as subtracting $3\frac{3}{5}$ all at once.

2. Fill in the blank to make the statement true.

   In order to add fourths and thirds, I need a common unit. I can rename both fractions as twelfths.

   a. $3\frac{1}{4} + 2\frac{2}{3} + \underline{\mathbf{3\frac{1}{12}}} = 9$

   $\mathbf{3\frac{3}{12} + 2\frac{8}{12} + \_\_\_\_ = 9}$

   $\mathbf{5\frac{11}{12} + \_\_\_\_ = 9}$

   $\mathbf{5\frac{11}{12} + 3\frac{1}{12} = 9}$

   I could solve this by subtracting $5\frac{11}{12}$ from 9, but I'm going to count on from $5\frac{11}{12}$ instead.

   $5\frac{11}{12}$ needs $\frac{1}{12}$ more to make 6. And then, 6 needs 3 more to make 9. So, $5\frac{11}{12} + 3\frac{1}{12} = 9$.

   $5\frac{11}{12} \xrightarrow{+\frac{1}{12}} 6 \xrightarrow{+3} 9$

When I look at this equation, I think, "There is *some number* that, when I subtract $2\frac{1}{2}$ and 15 from it, there is still $17\frac{1}{4}$ remaining." This helps me to visualize a tape diagram like this:

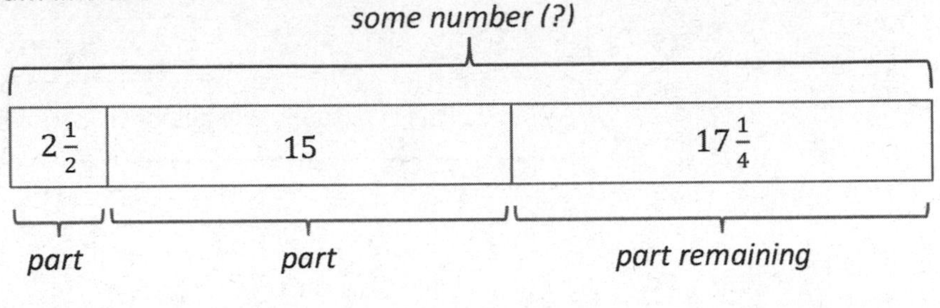

b. 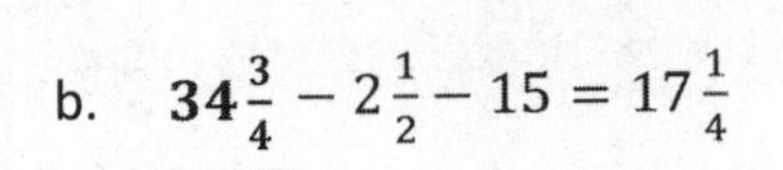 $\mathbf{34\frac{3}{4}} - 2\frac{1}{2} - 15 = 17\frac{1}{4}$

Therefore, if I add together these 3 parts, I can find out what that missing number is.

$$\mathbf{2\frac{1}{2} + 15 + 17\frac{1}{4}}$$
$$= \mathbf{34 + \left(\frac{1}{2} + \frac{1}{4}\right)}$$
$$= \mathbf{34\frac{3}{4}}$$

I can add the whole numbers and then add the fractions.

I can rename $\frac{1}{2}$ as $\frac{2}{4}$ in my head in order to add like units.

Name ______________________________ Date ________________

1. Rearrange the terms so that you can add or subtract mentally. Then, solve.

a. $1\frac{3}{4}+\frac{1}{2}+\frac{1}{4}+\frac{1}{2}$

b. $3\frac{1}{6}-\frac{3}{4}+\frac{5}{6}$

c. $5\frac{5}{8}-2\frac{6}{7}-\frac{2}{7}-\frac{5}{8}$

d. $\frac{7}{9}+\frac{1}{2}-\frac{3}{2}+\frac{2}{9}$

2. Fill in the blank to make the statement true.

a. $7\frac{3}{4}-1\frac{2}{7}-\frac{3}{2}=$ ________

b. $9\frac{5}{8}+1\frac{1}{4}+$ ________ $=14$

c. $\frac{7}{10} - \_\_\_\_ + \frac{3}{2} = \frac{6}{5}$

d. $\_\_\_\_ - 20 - 3\frac{1}{4} = 14\frac{5}{8}$

e. $\frac{17}{3} + \_\_\_\_ + \frac{5}{2} = 10\frac{4}{5}$

f. $23.1 + 1\frac{7}{10} - \_\_\_\_ = \frac{66}{10}$

3. Laura bought $8\frac{3}{10}$ yd of ribbon. She used $1\frac{2}{5}$ yd to tie a package and $2\frac{1}{3}$ yd to make a bow. Joe later gave her $4\frac{3}{5}$ yd. How much ribbon does she now have?

4. Mia bought $10\frac{1}{9}$ lb of flour. She used $2\frac{3}{4}$ lb of flour to bake banana cakes and some to bake chocolate cakes. After baking all the cakes, she had $3\frac{5}{6}$ lb of flour left. How much flour did she use to bake the chocolate cakes?

1. Nikki bought 10 meters of cloth. She used $2\frac{1}{4}$ meters for a dress and $1\frac{3}{5}$ meters for a shirt. How much cloth did she have left?

There are different ways to solve this problem. I could subtract the length of the dress and the shirt from the total length of the cloth.

I'll draw a tape diagram and label the whole as 10 m and the parts as $2\frac{1}{4}$ m and $1\frac{3}{5}$ m.

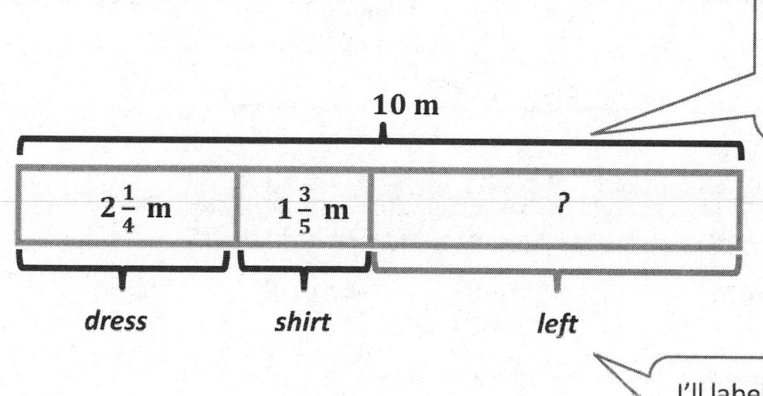

I'll label the part that's left with a question mark because that's what I'm trying to find.

I can subtract the whole numbers first.
$10 - 2 - 1 = 7$

$$10 - 2\frac{1}{4} - 1\frac{3}{5}$$
$$= 7 - \frac{1}{4} - \frac{3}{5}$$
$$= 7 - \frac{5}{20} - \frac{12}{20}$$
$$= 6\frac{20}{20} - \frac{5}{20} - \frac{12}{20}$$
$$= 6\frac{3}{20}$$

I can rename these fractions as twentieths in order to subtract.
$\frac{1}{4} = \frac{5}{20}$, and $\frac{3}{5} = \frac{12}{20}$.

I need to rename 7 as $6\frac{20}{20}$ so I can subtract.

She had $6\frac{3}{20}$ meters of cloth left.

2. Jose bought $3\frac{1}{5}$ kg of carrots, $1\frac{3}{4}$ kg of potatoes, and $2\frac{2}{5}$ kg of broccoli. What's the total weight of the vegetables?

I'll use addition to find the total weight of the vegetables.

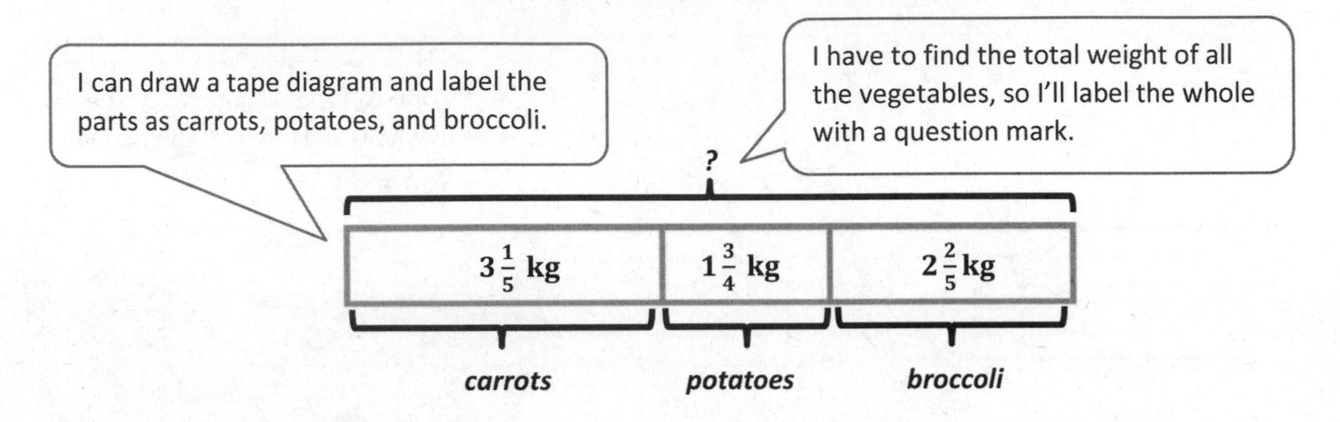

I can add the whole numbers.

$3 + 1 + 2 = 6$

$$3\frac{1}{5} + 1\frac{3}{4} + 2\frac{2}{5}$$

$$= 6 + \frac{1}{5} + \frac{3}{4} + \frac{2}{5}$$

$$= 6 + \frac{4}{20} + \frac{15}{20} + \frac{8}{20}$$

I need to rename the fractions with a common unit of twentieths.

$\frac{1}{5} = \frac{4}{20}$, $\frac{3}{4} = \frac{15}{20}$, and $\frac{2}{5} = \frac{8}{20}$.

$$= 6 + \frac{27}{20}$$

$$= 6 + \frac{20}{20} + \frac{7}{20}$$

$\frac{27}{20} = \frac{20}{20} + \frac{7}{20} = 1\frac{7}{20}$

$$= 7\frac{7}{20}$$

The total weight of the vegetables is $7\frac{7}{20}$ kilograms.

Name ______________________________ Date ________________

Solve the word problems using the RDW strategy. Show all of your work.

1. A baker buys a 5 lb bag of sugar. She uses $1\frac{2}{3}$ lb to make some muffins and $2\frac{3}{4}$ lb to make a cake. How much sugar does she have left?

2. A boxer needs to lose $3\frac{1}{2}$ kg in a month to be able to compete as a flyweight. In three weeks, he lowers his weight from 55.5 kg to 53.8 kg. How many kilograms must the boxer lose in the final week to be able to compete as a flyweight?

3. A construction company builds a new rail line from Town A to Town B. They complete $1\frac{1}{4}$ miles in their first week of work and $1\frac{2}{3}$ miles in the second week. If they still have $25\frac{3}{4}$ miles left to build, what is the distance from Town A to Town B?

4. A catering company needs 8.75 lb of shrimp for a small party. They buy $3\frac{2}{3}$ lb of jumbo shrimp, $2\frac{5}{8}$ lb of medium-sized shrimp, and some mini-shrimp. How many pounds of mini-shrimp do they buy?

5. Mark breaks up a 9-hour drive into 3 segments. He drives $2\frac{1}{2}$ hours before stopping for lunch. After driving some more, he stops for gas. If the second segment of his drive was $1\frac{2}{3}$ hours longer than the first segment, how long did he drive after stopping for gas?

I know $\frac{1}{4}$ plus $\frac{3}{4}$ is equal to $\frac{4}{4}$, or 1.

Draw the following ribbons.

a. 1 ribbon. The piece shown below is only $\frac{1}{4}$ of the whole. Complete the drawing to show the whole ribbon.

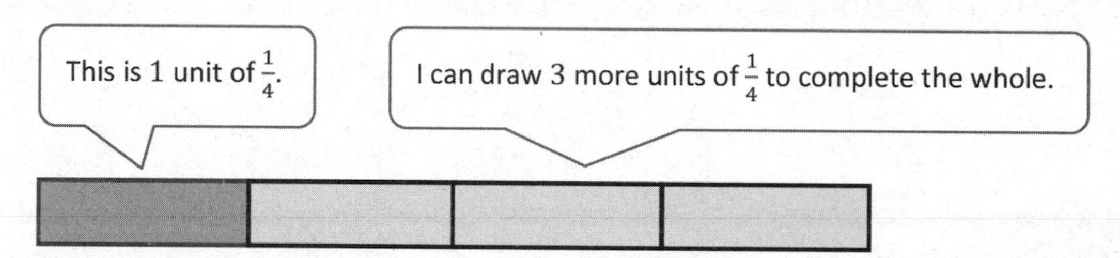

b. 1 ribbon. The piece shown below is $\frac{3}{5}$ of the whole. Complete the drawing to show the whole ribbon.

I know $\frac{3}{5}$ plus $\frac{2}{5}$ is equal to $\frac{5}{5}$, or 1.

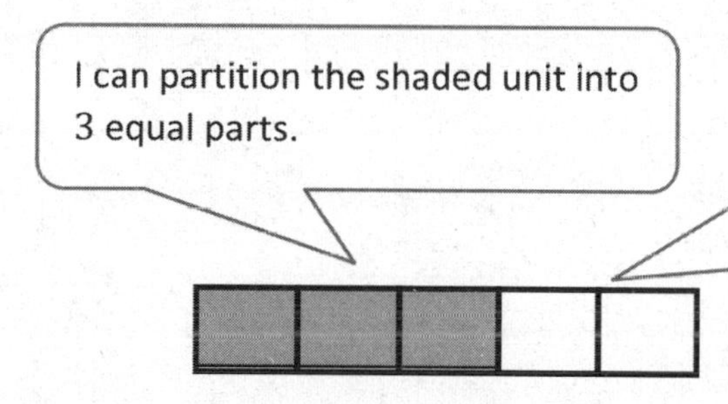

I need to draw 2 more units to make a total of 5 parts. Now, the shaded part represents $\frac{3}{5}$, and the unshaded part represents $\frac{2}{5}$.

c. 2 ribbons, A and B. One sixth of A is equal to all of B. Draw a picture of the ribbons.

I know that ribbon A must be longer than B. More specifically, ribbon B is just 1 sixth of A. This also means that ribbon A is 6 times longer than ribbon B.

I can draw one large unit to represent ribbon A. Then, I can partition it into 6 equal parts.

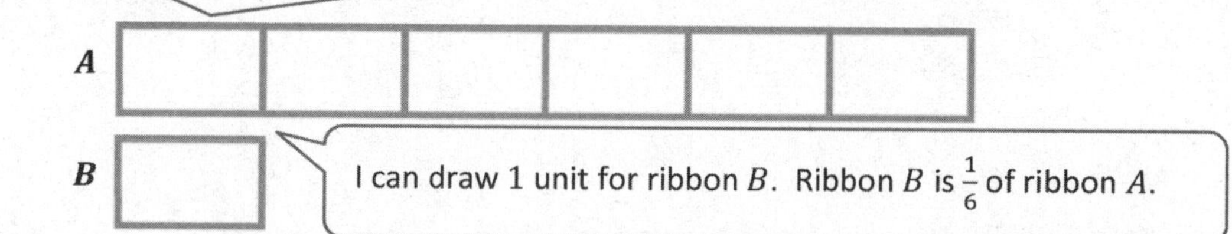

Name ______________________________ Date ________________

Draw the following roads.

a. 1 road. The piece shown below is only $\frac{3}{7}$ of the whole. Complete the drawing to show the whole road.

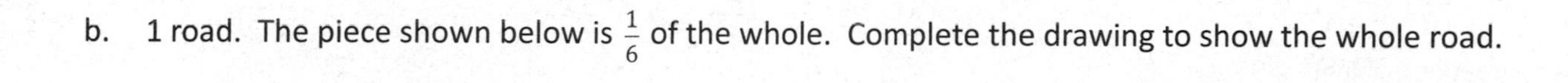

b. 1 road. The piece shown below is $\frac{1}{6}$ of the whole. Complete the drawing to show the whole road.

c. 3 roads, A, B, and C. B is three times longer than A. C is twice as long as B. Draw the roads. What fraction of the total length of the roads is the length of A? If Road B is 7 miles longer than Road A, what is the length of Road C?

d. Write your own road problem with 2 or 3 lengths.

Grade 5
Module 4

1. A group of students measured the height of their bean sprout to the nearest quarter inch. Draw a line plot to represent their data:

$2\frac{1}{2}$, $1\frac{1}{4}$, 2, $3\frac{1}{2}$, $2\frac{1}{4}$, 2, $2\frac{1}{2}$, 2, $2\frac{1}{2}$, $2\frac{1}{4}$, $3\frac{1}{4}$

I can put an X above the number line for each measurement in this set of data.

Bean Sprout Height

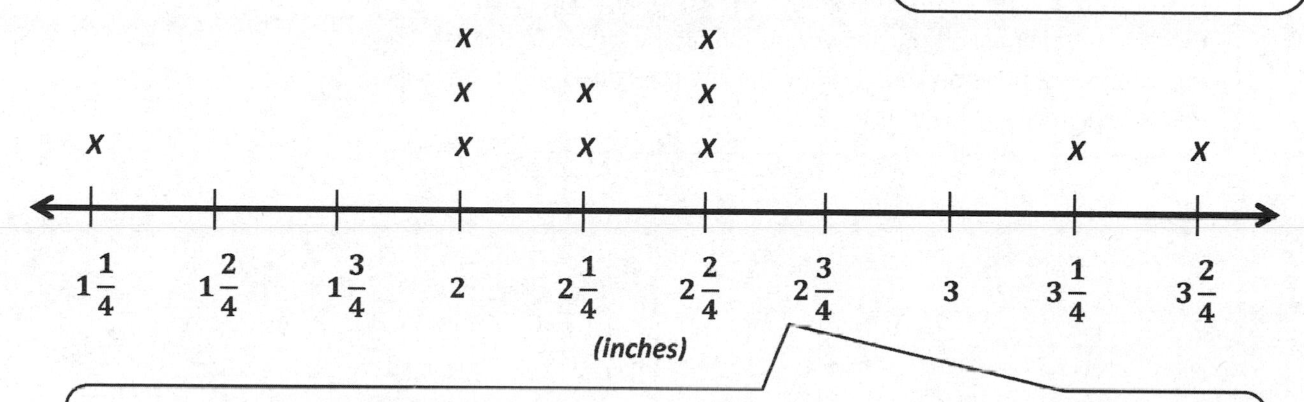

Since the data set includes values of both half, quarter, and whole inches, I can draw a number line that shows values between $1\frac{1}{4}$ and $3\frac{2}{4}$ and all of the $\frac{1}{4}$ inches in between.

2. Answer the following questions.

Once my line plot is created, I can use it to help me answer these questions.

a. Which bean sprout is the tallest?

The tallest sprout is $3\frac{1}{2}$ inches.

b. Which bean sprout is the shortest?

$1\frac{1}{4}$ inches.

c. Which measurement is the most frequent?

The most frequent values are 2 inches and $2\frac{1}{2}$ inches.

Most frequent means the value listed the most times. Since both 2 and $2\frac{1}{2}$ were listed three times, both values are considered most frequent.

d. What is the total height of all the bean sprouts?

The total height of all the values is 26 inches.

I made sure to add all eleven values. For example, I had to add 2 three times. I checked my answer by adding the values in the list and then the values on the number line to make sure both sums were the same.

Name ____________________________________ Date ____________________

A meteorologist set up rain gauges at various locations around a city and recorded the rainfall amounts in the table below. Use the data in the table to create a line plot using $\frac{1}{8}$ inches.

Location	Rainfall Amount (inches)
1	$\frac{1}{8}$
2	$\frac{3}{8}$
3	$\frac{3}{4}$
4	$\frac{3}{4}$
5	$\frac{1}{4}$
6	$1\frac{1}{4}$
7	$\frac{1}{8}$
8	$\frac{1}{4}$
9	1
10	$\frac{1}{8}$

a. Which location received the most rainfall?

b. Which location received the least rainfall?

c. Which rainfall measurement was the most frequent?

d. What is the total rainfall in inches?

1. Draw a picture to show the division. Express your answer as a fraction.

   a. $1 \div 3 = \mathbf{3\ \textit{thirds} \div 3 = 1\ \textit{third}} = \frac{1}{3}$

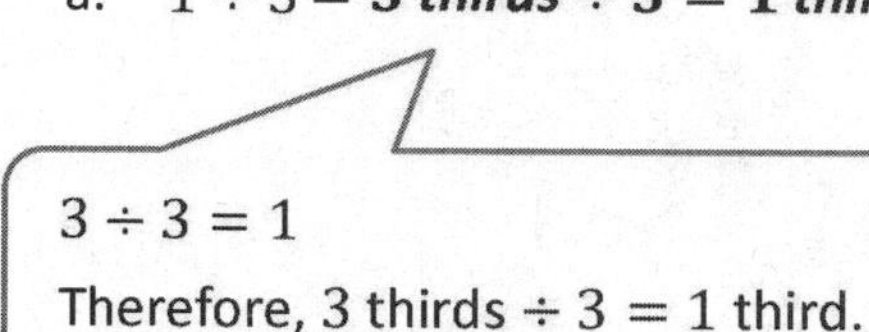

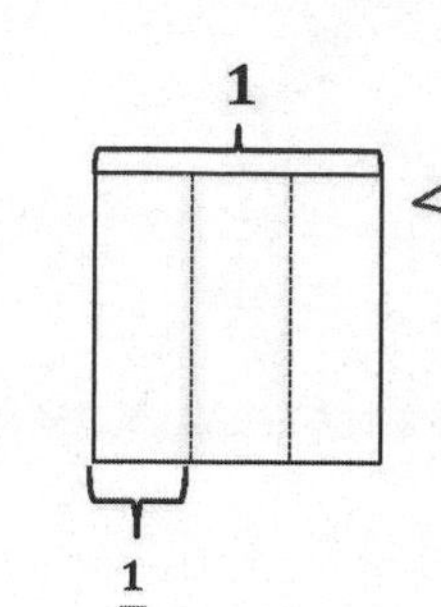

I can think about $1 \div 3$ as 1 cracker being shared equally by 3 people. Each person gets $\frac{1}{3}$ of the cracker.

   b. $2 \div 5 = \mathbf{10\ \textit{fifths} \div 5 = 2\ \textit{fifths}} = \frac{2}{5}$

$10 \div 5 = 2$
Therefore, 10 fifths $\div$ 5 = 2 fifths.

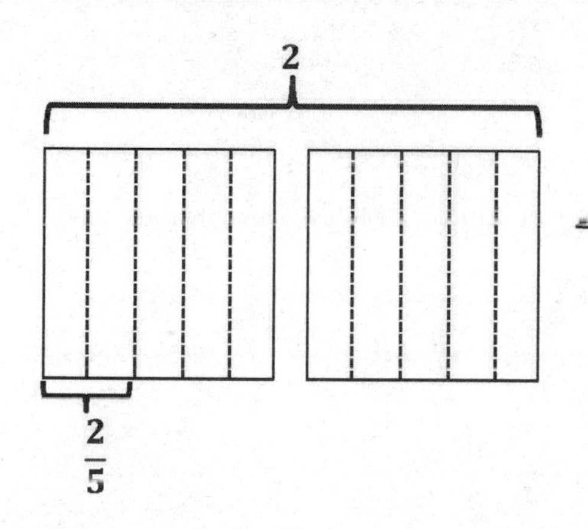

If 2 crackers were shared equally by 5 people, each person would get $\frac{2}{5}$ of a cracker.

2. Fill in the blanks to make true number sentences.

   a. $15 \div 4 = \frac{\mathbf{15}}{\mathbf{4}}$

   I can write a division expression as a fraction.

   b. $\frac{5}{3} = \underline{\mathbf{5}} \div \underline{\mathbf{3}}$

   I can interpret a fraction as a division expression.

   c. $2\frac{1}{2} = \underline{\mathbf{5}} \div \underline{\mathbf{2}}$

   I can express this mixed number as a fraction greater than 1.

   $2\frac{1}{2} = \frac{5}{2}$

   If 5 crackers were shared equally by 2 people, each person would get 5 halves, or $2\frac{1}{2}$ crackers.

Name ______________________________ Date ________________

1. Draw a picture to show the division. Express your answer as a fraction.

 a. $1 \div 4$

 b. $3 \div 5$

 c. $7 \div 4$

2. Using a picture, show how six people could share four sandwiches. Then, write an equation and solve.

3. Fill in the blanks to make true number sentences.

a. $2 \div 7 = \frac{\quad}{\quad}$

b. $39 \div 5 = \frac{\quad}{\quad}$

c. $13 \div 3 = \frac{\quad}{\quad}$

d. $\frac{9}{5}$ = ______ ÷ ______

e. $\frac{19}{28}$ = ______ ÷ ______

f. $1\frac{3}{5}$ = ______ ÷ ______

1. Fill in the chart.

Division Expression	Unit Form	Improper Fraction	Mixed Number	Standard Algorithm (Write your answer in whole numbers and fractional units. Then check.)
a. $3 \div 2$	***6 halves*** $\div 2 =$ ***3 halves***	$\frac{3}{2}$	$1\frac{1}{2}$	$1\frac{1}{2}$; $2\overline{)3}$, -2, 1. ***Check:*** $2 \times 1\frac{1}{2} = 1\frac{1}{2} + 1\frac{1}{2} = 3$

I can visualize the drawings I made in the previous lesson. 3 crackers are shared equally by 2 people. I could partition each cracker into 2 equal parts and then share the 6 halves.

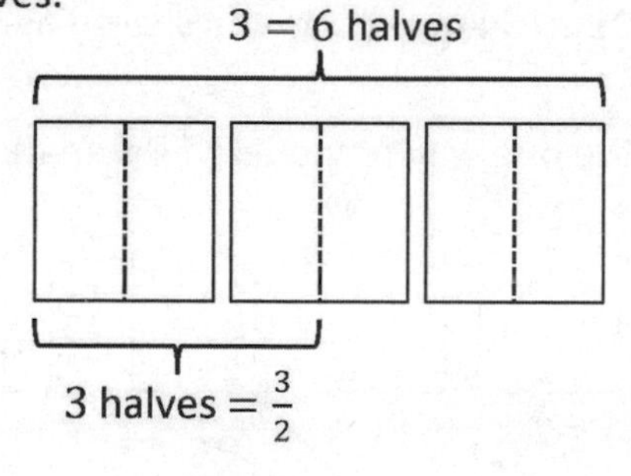

I can think of this another way too. Since there are 3 crackers being shared equally by 2 people, each person could get 1 whole cracker and $\frac{1}{2}$ of another.

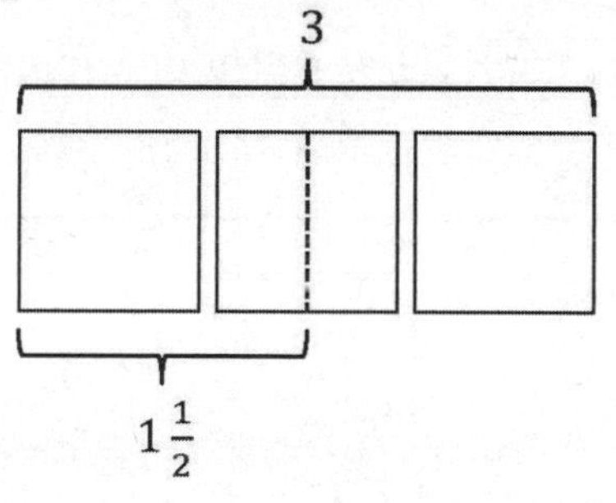

Division Expression	Unit Form	Improper Fraction	Mixed Numbers	Standard Algorithm (Write your answer in whole numbers and fractional units. Then check.)
b. $5 \div 3$	15 *thirds* $\div 3 =$ 5 *thirds*	$\frac{5}{3}$	$1\frac{2}{3}$	$3 \overline{)5}$ → $1\frac{2}{3}$; -3; 2 *Check:* $3 \times 1\frac{2}{3}$ $= 1\frac{2}{3} + 1\frac{2}{3} + 1\frac{2}{3}$ $= 3\frac{6}{3}$ $= 3 + 2$ $= 5$

This time I am given the mixed number. I know that $1\frac{2}{3}$ is the same as $\frac{3}{3} + \frac{2}{3}$, which is equal to $\frac{5}{3}$.

I can think of $\frac{5}{3}$ as a division expression, $5 \div 3$.

The standard algorithm makes sense. If there were 5 crackers being shared equally by 3 people, each person could get 1 whole cracker, and then the remaining 2 crackers would be partitioned into 3 equal parts and shared as thirds.

I can visualize one way to model this scenario:

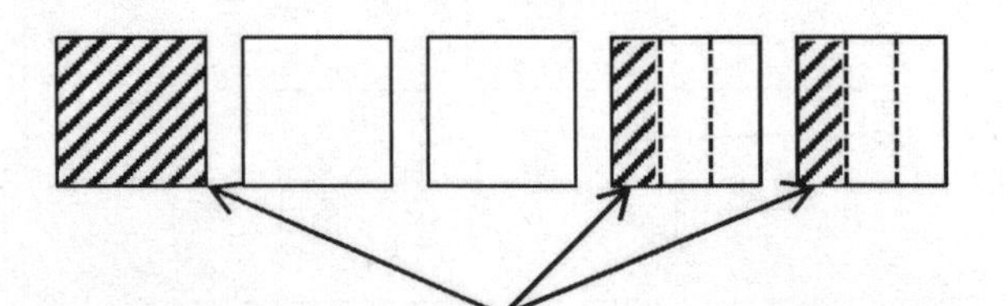

Each person gets 1 whole cracker and $\frac{2}{3}$ of a cracker.

Name ______________________________ Date ________________

1. Fill in the chart. The first one is done for you.

Division Expression	Unit Forms	Improper Fractions	Mixed Numbers	Standard Algorithm (Write your answer in whole numbers and fractional units. Then check.)
a. $4 \div 3$	12 thirds ÷ 3 = 4 thirds	$\frac{4}{3}$	$1\frac{1}{3}$	$3\overline{)4}$ quotient $1\frac{1}{3}$; -3; 1 Check $3 \times 1\frac{1}{3} = 1\frac{1}{3} + 1\frac{1}{3} + 1\frac{1}{3}$ $= 3 + \frac{3}{3}$ $= 3 + 1$ $= 4$
b. ___ ÷ ___	___ fifths ÷ 5 = ___ fifths		$1\frac{2}{5}$	
c. ___ ÷ ___	___ halves ÷ 2 = ___ halves			$2\overline{)7}$
d. $7 \div 4$		$\frac{7}{4}$		

2. A coffee shop uses 4 liters of milk every day.

   a. If there are 15 liters of milk in the refrigerator, after how many days will more milk need to be purchased? Explain how you know.

   b. If only half as much milk is used each day, after how many days will more milk need to be purchased?

3. Polly buys 14 cupcakes for a party. The bakery puts them into boxes that hold 4 cupcakes each.

   a. How many boxes will be needed for Polly to bring all the cupcakes to the party? Explain how you know.

   b. If the bakery completely fills as many boxes as possible, what fraction of the last box is empty? How many more cupcakes are needed to fill this box?

Draw a tape diagram to solve. Express your answer as a fraction. Show the addition sentence to support your answer.

$5 \div 4 = \frac{5}{4} = 1\frac{1}{4}$

I can model $5 \div 4$ by drawing a tape diagram. The whole tape represents the dividend, 5. The divisor is 4, so I partition the model into 4 equal parts, or units.

I can think of the expression $5 \div 4$ as 5 crackers being shared equally by 4 people. This unit here represents how much 1 person gets.

5

?

Now that I've divided, I know that each of these four units has a value of $1\frac{1}{4}$.

4 *units* = 5

1 *unit* = $5 \div 4 = \frac{5}{4} = 1\frac{1}{4}$

My tape diagram shows me that the 4 parts, or units, are equal to 5. So, I can find the value of 1 unit by dividing, $5 \div 4$.

$$\begin{array}{r|l} & 1\frac{1}{4} \\ 4 & 5 \\ & 4 \\ \hline & 1 \end{array}$$

Check:

$4 \times 1\frac{1}{4}$

$= 1\frac{1}{4} + 1\frac{1}{4} + 1\frac{1}{4} + 1\frac{1}{4}$

$= 4 + \frac{4}{4}$

$= 5$

Name ______________________________ Date ______________

1. Draw a tape diagram to solve. Express your answer as a fraction. Show the addition sentence to support your answer. The first one is done for you.

a. $1 \div 4 = \frac{1}{4}$

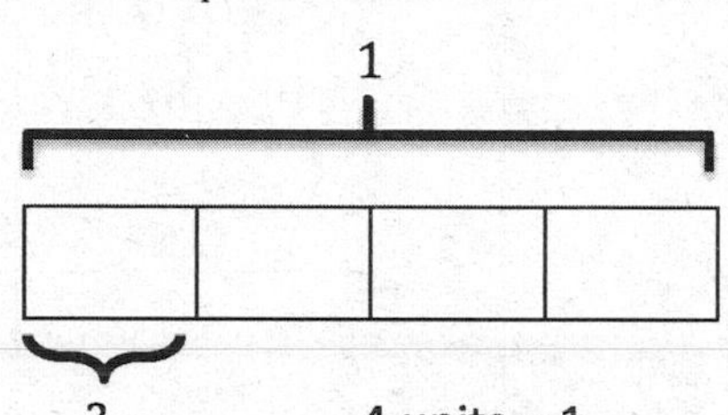

4 units = 1

1 unit = $1 \div 4$

$= \frac{1}{4}$

$$\begin{array}{r} 0\ \tfrac{1}{4} \\ 4 \overline{)\ 1} \\ -\ 0 \\ \hline 1 \end{array}$$

Check: $4 \times \frac{1}{4}$

$= \frac{1}{4} + \frac{1}{4} + \frac{1}{4} + \frac{1}{4}$

$= \frac{4}{4}$

$= 1$

b. $4 \div 5 =$ ——

c. $8 \div 5 =$ ——

d. $14 \div 3 =$ ——

2. Fill in the chart. The first one is done for you.

Division Expression	Fraction	Between which two whole numbers is your answer?	Standard Algorithm
a. $16 \div 5$	$\frac{16}{5}$	3 and 4	$3\frac{1}{5}$ 5 ⟌ 16 -15 1
b. ____ ÷ ____	$\frac{3}{4}$	0 and 1	
c. ____ ÷ ____	$\frac{7}{2}$		2 ⟌ 7
d. ____ ÷ ____	$\frac{81}{90}$		

3. Jackie cut a 2-yard spool into 5 equal lengths of ribbon.

   a. What is the length of each ribbon in yards? Draw a tape diagram to show your thinking.

   b. What is the length of each ribbon in feet? Draw a tape diagram to show your thinking.

4. Baa Baa, the black sheep, had 7 pounds of wool. If he separated the wool equally into 3 bags, how much wool would be in 2 bags?

5. An adult sweater is made from 2 pounds of wool. This is 3 times as much wool as it takes to make a baby sweater. How much wool does it take to make a baby sweater? Use a tape diagram to solve.

Kenneth divided 15 cups of whole wheat flour equally to make 4 loaves of bread

a. How much whole wheat flour went into each loaf?

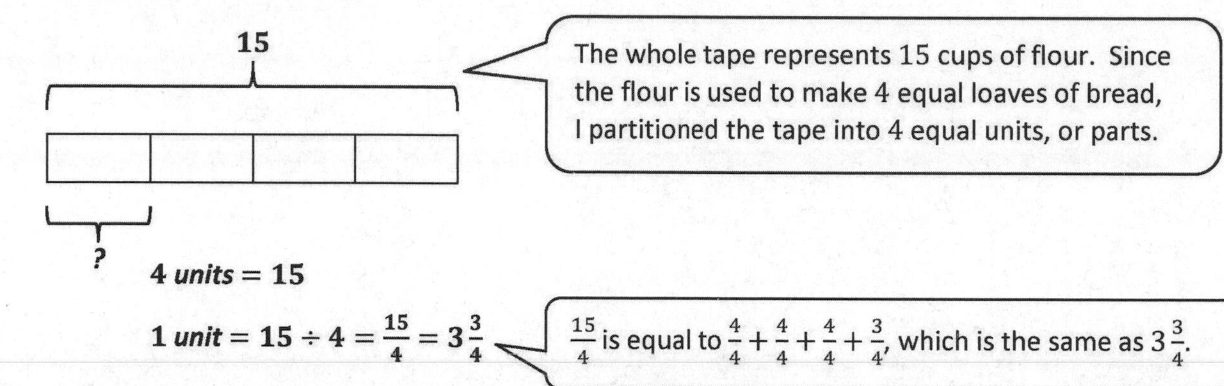

Kenneth used $3\frac{3}{4}$ cups of whole wheat flour for each loaf of bread.

b. How many cups of whole wheat flour are in 3 loaves of bread?

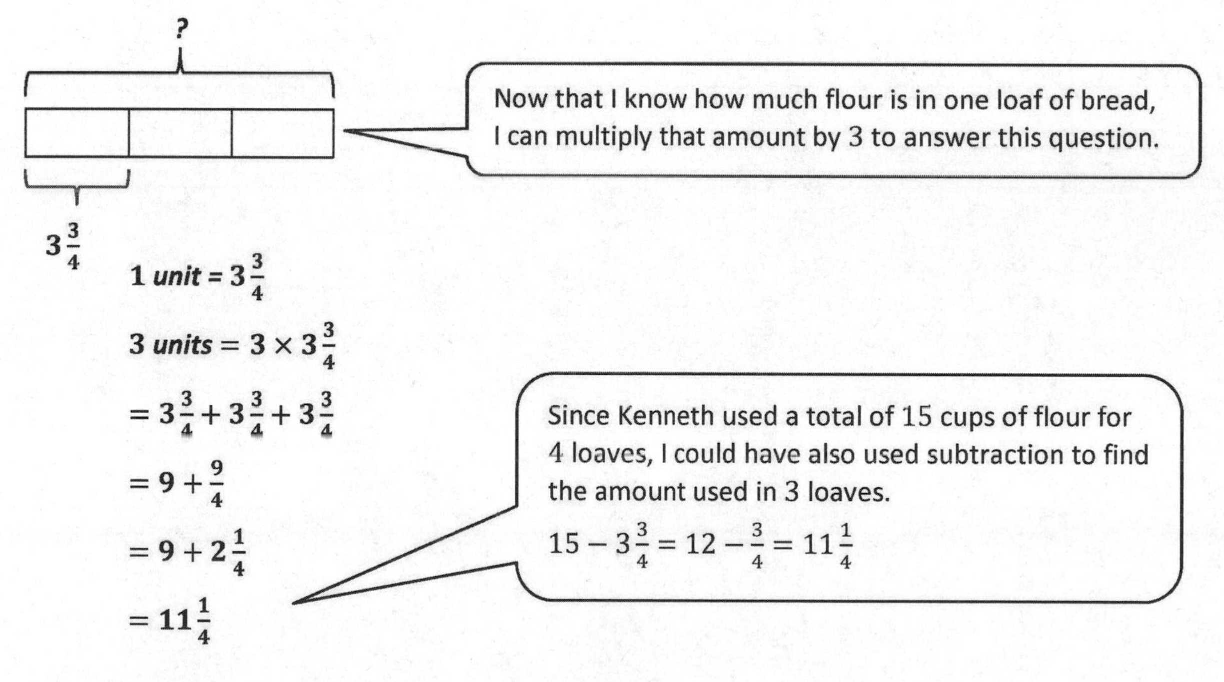

There are $11\frac{1}{4}$ cups of whole wheat flour in 3 loaves.

Name ______________________________ Date ______________

1. When someone donated 14 gallons of paint to Rosendale Elementary School, the fifth grade decided to use it to paint murals. They split the gallons equally among the four classes.

   a. How much paint did each class have to paint their mural?

   b. How much paint will three classes use? Show your thinking using words, numbers, or pictures.

   c. If 4 students share a 30-square-foot wall equally, how many square feet of the wall will be painted by each student?

   d. What fraction of the wall will each student paint?

2. Craig bought a 3-foot-long baguette and then made 4 equally sized sandwiches with it.

   a. What portion of the baguette was used for each sandwich? Draw a visual model to help you solve this problem.

   b. How long, in feet, is one of Craig's sandwiches?

   c. How many inches long is one of Craig's sandwiches?

3. Scott has 6 days to save enough money for a $45 concert ticket. If he saves the same amount each day, what is the minimum amount he must save each day in order to reach his goal? Express your answer in dollars.

1. Find the value of the following.

```
*  |  *  |  *
*  |  *  |  *
*  |  *  |  *
*  |  *  |  *
*  |  *  |  *
```

> The array shows a total of 15 stars. Each column represents 1 third.

$\frac{1}{3}$ ***of*** **15 = 5**

$\frac{2}{3}$ ***of*** **15 = 10**

> To find 2 thirds, I can count the number of stars in two columns.

$\frac{3}{3}$ ***of*** **15 = 15**

> $\frac{3}{3}$ represents *all* of the stars, or the amount found in all 3 columns.

2. Find $\frac{3}{4}$ of 12. Draw a set, and shade to show your thinking.

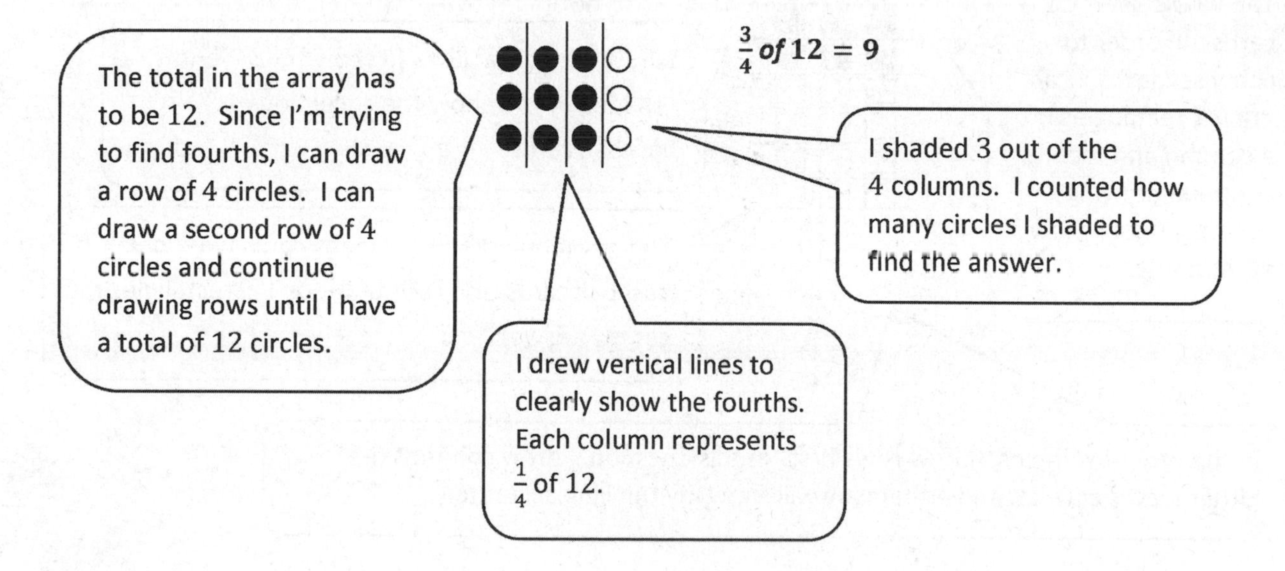

3. How does knowing $\frac{1}{3}$ of 18 help you find $\frac{2}{3}$ of 18? Draw a picture to explain your thinking.

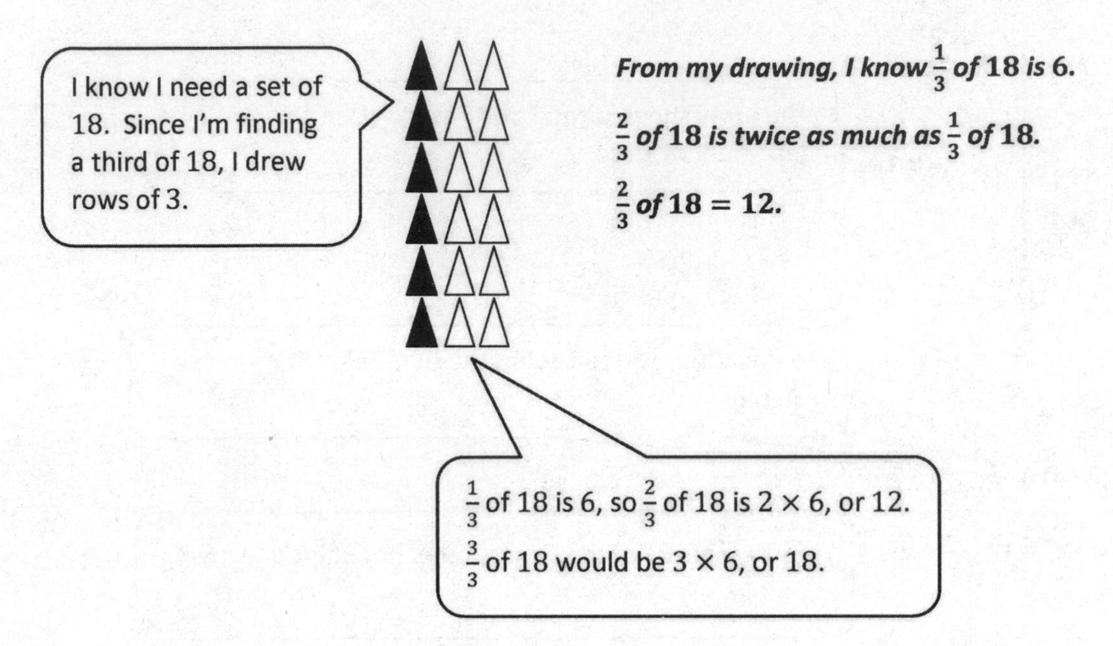

4. Michael collected 21 sports cards. $\frac{3}{7}$ of the cards are baseball cards. How many cards are not baseball cards?

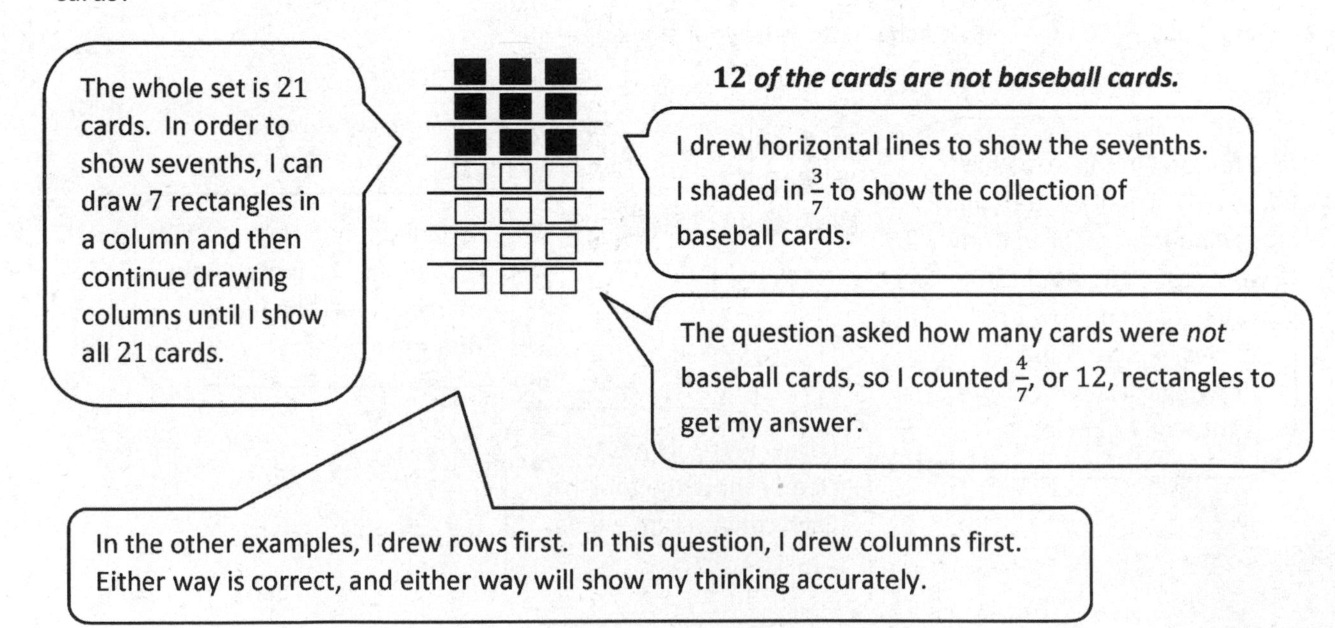

EUREKA MATH

Name ______________________________ Date ________________

1. Find the value of each of the following.

a.

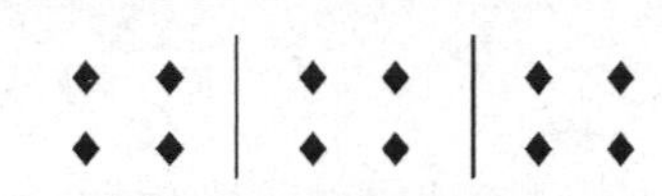

$\frac{1}{3}$ of 12 =

$\frac{2}{3}$ of 12 =

$\frac{3}{3}$ of 12 =

b.

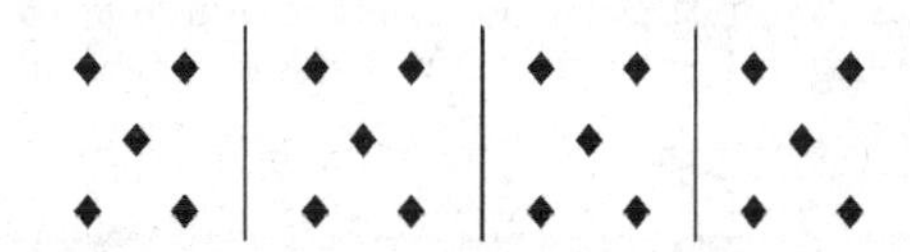

$\frac{1}{4}$ of 20 =

$\frac{2}{4}$ of 20 =

$\frac{3}{4}$ of 12 =

$\frac{4}{4}$ of 12 =

c.

$\frac{1}{5}$ of 35 =

$\frac{2}{5}$ of 35 =

$\frac{3}{5}$ of 35 =

$\frac{4}{5}$ of 35 =

$\frac{5}{5}$ of 35 =

$\frac{6}{5}$ of 35 =

2. Find $\frac{2}{3}$ of 18. Draw a set and shade to show your thinking.

3. How does knowing $\frac{1}{5}$ of 10 help you find $\frac{3}{5}$ of 10? Draw a picture to explain your thinking.

4. Sara just turned 18 years old. She spent $\frac{4}{9}$ of her life living in Rochester, NY. How many years did Sara live in Rochester?

5. A farmer collected 12 dozen eggs from her chickens. She sold $\frac{5}{6}$ of the eggs at the farmers' market and gave the rest to friends and neighbors.

   a. How many dozen eggs did the farmer give away? How many eggs did she give away?

   b. She sold each dozen for $4.50. How much did she earn from the eggs she sold?

I can draw a tape diagram and label the whole as 25. I need to find fifths, so I partition the whole into five units, or parts.

Solve using a tape diagram.

a. $\frac{1}{5}$ of $25 = \mathbf{5}$

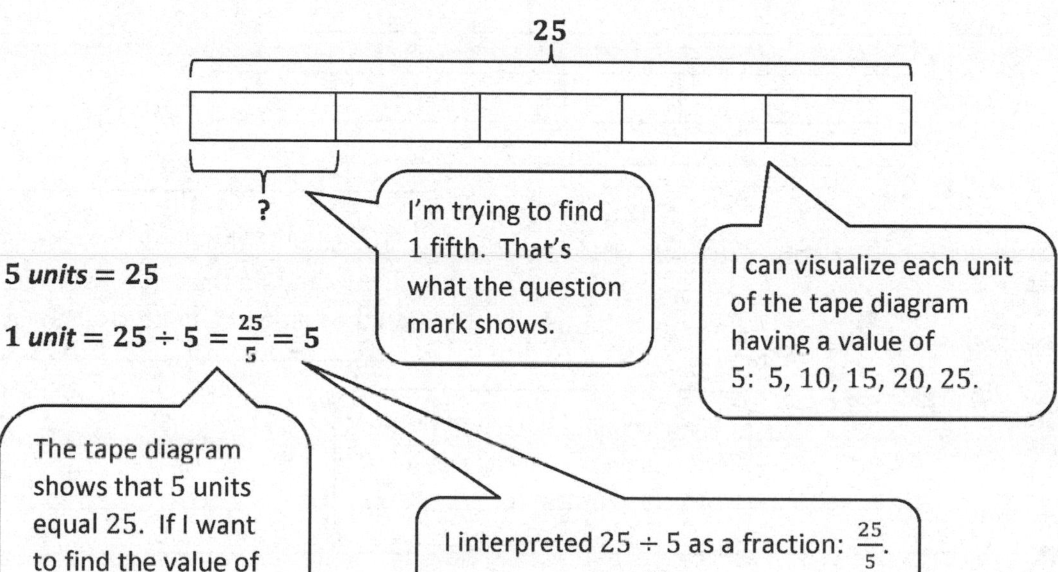

$\mathbf{5\ \textit{units} = 25}$

$\mathbf{1\ \textit{unit} = 25 \div 5 = \frac{25}{5} = 5}$

The tape diagram shows that 5 units equal 25. If I want to find the value of 1 unit, I need to divide 25 by 5.

I interpreted $25 \div 5$ as a fraction: $\frac{25}{5}$. Then I simplified $\frac{25}{5}$ as 5.

b. $\frac{3}{4} \times 16 = \mathbf{12}$

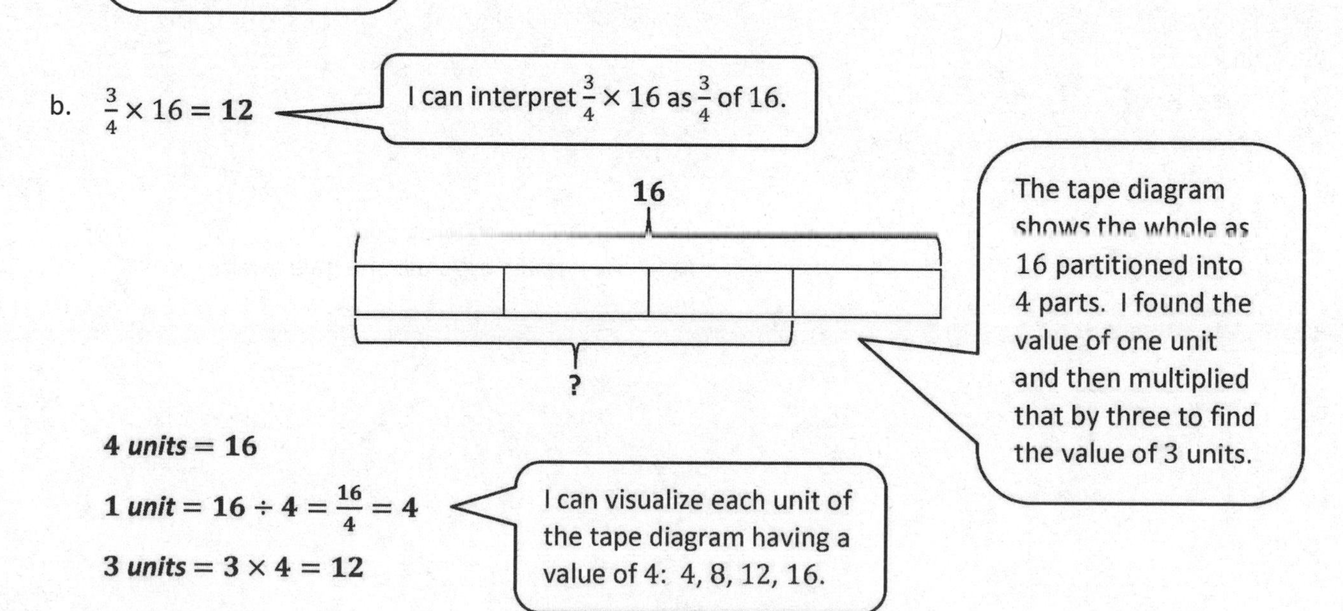

$\mathbf{4\ \textit{units} = 16}$

$\mathbf{1\ \textit{unit} = 16 \div 4 = \frac{16}{4} = 4}$

$\mathbf{3\ \textit{units} = 3 \times 4 = 12}$

I can visualize each unit of the tape diagram having a value of 4: 4, 8, 12, 16.

I can interpret this as $\frac{5}{6}$ of ? = 25.

c. $\frac{5}{6}$ of a number is 25. What's the number?

In this problem, I am given the value of some parts, and I need to find the value of the whole.

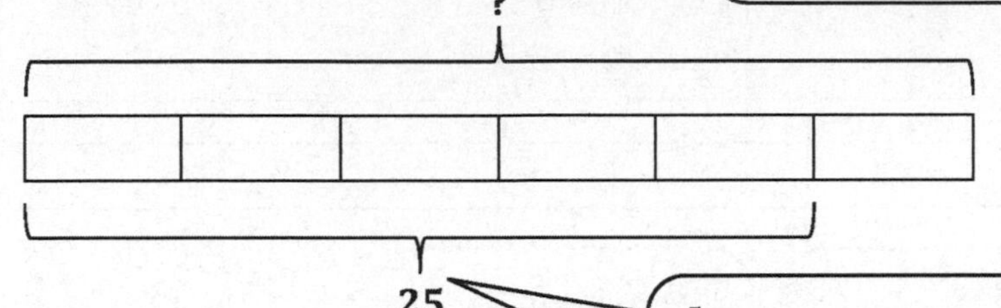

$\frac{5}{6} = 25$, so these 5 units have a value of 25. If I can find the value of 1 unit, I can find the value of 6 units, or the whole.

5 *units* = 25

1 *unit* = $25 \div 5 = \frac{25}{5} = 5$

6 *units* = $6 \times 5 = 30$

I can visualize each unit of the tape diagram having a value of 5: 5, 10, 15, 20, 25, 30.

***The number is* 30.**

Name ______________________________ Date ________________

1. Solve using a tape diagram.

a. $\frac{1}{4}$ of 24

b. $\frac{1}{4}$ of 48

c. $\frac{2}{3} \times 18$

d. $\frac{2}{6} \times 18$

e. $\frac{3}{7} \times 49$

f. $\frac{3}{10} \times 120$

g. $\frac{1}{3} \times 31$

h. $\frac{2}{5} \times 20$

i. $\frac{1}{4} \times 25$

j. $\frac{3}{5} \times 25$

k. $\frac{3}{4}$ of a number is 27. What's the number?

l. $\frac{2}{5}$ of a number is 14. What's the number?

2. Solve using tape diagrams.

   a. A skating rink sold 66 tickets. Of these, $\frac{2}{3}$ were children's tickets, and the rest were adult tickets. What total number of adult tickets were sold?

   b. A straight angle is split into two smaller angles as shown. The smaller angle's measure is $\frac{1}{6}$ that of a straight angle. What is the value of angle a?

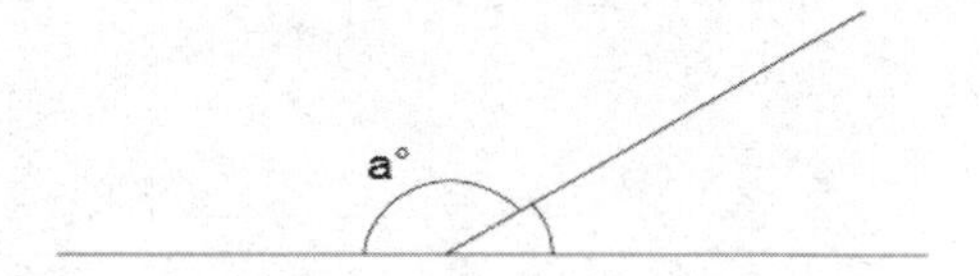

   c. Annabel and Eric made 17 ounces of pizza dough. They used $\frac{5}{8}$ of the dough to make a pizza and used the rest to make calzones. What is the difference between the amount of dough they used to make pizza and the amount of dough they used to make calzones?

   d. The New York Rangers hockey team won $\frac{3}{4}$ of their games last season. If they lost 21 games, how many games did they play in the entire season?

1. Rewrite the following expressions as shown in the example.

   Example: $\frac{4}{7}+\frac{4}{7}+\frac{4}{7}=\frac{3\times 4}{7}=\frac{12}{7}$

   This expression is repeatedly adding 2 fifths. I can write it as a multiplication expression. This is the same as $4\times\frac{2}{5}$, or $\frac{4\times 2}{5}$.

   a. $\frac{3}{2}+\frac{3}{2}+\frac{3}{2}$

   $\frac{3}{2}+\frac{3}{2}+\frac{3}{2}=\frac{3\times 3}{2}=\frac{9}{2}$

   b. $\frac{2}{5}+\frac{2}{5}+\frac{2}{5}+\frac{2}{5}$

   $\mathbf{\frac{2}{5}+\frac{2}{5}+\frac{2}{5}+\frac{2}{5}=\frac{4\times 2}{5}=\frac{8}{5}}$

2. Solve each problem in two different ways. Express your answer in simplest form.

   a. $\frac{2}{5}\times 30$

   $\mathbf{\frac{2}{5}\times 30=\frac{2\times 30}{5}=\frac{60}{5}=12}$

   In this method, I simplified after I multiplied.

   $\mathbf{\frac{2}{5}\times 30=\frac{2\times \cancel{30}\,6}{\cancel{5}\,1}=12}$

   In this method, I see that 30 and 5 have a common factor of 5. I can divide both 30 and 5 by 5, and now I can think of the fraction as $\frac{2\times 6}{1}$.

   b. $32\times\frac{7}{8}$

   This method involved some larger numbers that are challenging to do mentally.

   $\mathbf{32\times\frac{7}{8}=\frac{32\times 7}{8}=\frac{224}{8}=28}$

   Dividing by a common factor of 8 made this method much simpler! I can do this mentally.

   $\mathbf{32\times\frac{7}{8}=\frac{4\,\cancel{32}\times 7}{\cancel{8}\,1}=28}$

3. Solve any way you choose.

   $\frac{3}{4}\times 60$

   $\mathbf{\frac{3}{4}\times 60=\frac{3\times 60}{4}=\frac{180}{4}=45}$

   Since there are 60 minutes in an hour, this is the expression I can use to find how many minutes are in $\frac{3}{4}$ of an hour.

   $\frac{3}{4}$ hour = __ minutes

   $\frac{3}{4}$ ***hour* = 45 *minutes***

   I could have solved by simplifying before I multiplied.

   $\frac{3}{4}\times 60=\frac{3\times \cancel{60}\,15}{\cancel{4}\,1}=45$

Name ______________________________ Date ______________

1. Rewrite the following expressions as shown in the example.

Example: $\frac{2}{3}+\frac{2}{3}+\frac{2}{3}+\frac{2}{3}=\frac{4\times 2}{3}=\frac{8}{3}$

a. $\frac{5}{3}+\frac{5}{3}+\frac{5}{3}$

b. $\frac{13}{5}+\frac{13}{5}$

c. $\frac{9}{4}+\frac{9}{4}+\frac{9}{4}$

2. Solve each problem in two different ways as modeled in the example.

Example: $\frac{2}{3}\times 6=\frac{2\times 6}{3}=\frac{12}{3}=4$ $\quad\frac{2}{3}\times 6=\frac{2\times \cancel{6}^{2}}{\cancel{3}_{1}}=4$

a. $\frac{3}{4}\times 16$ $\quad\frac{3}{4}\times 16$

b. $\frac{4}{3}\times 12$ $\quad\frac{4}{3}\times 12$

c. $40\times\frac{11}{10}$ $\quad 40\times\frac{11}{10}$

d. $\frac{7}{6}\times 36$ $\quad\frac{7}{6}\times 36$

e. $24\times\frac{5}{8}$ $\quad 24\times\frac{5}{8}$

f. $18 \times \frac{5}{12}$ $\qquad$ $18 \times \frac{5}{12}$

g. $\frac{10}{9} \times 21$ $\qquad$ $\frac{10}{9} \times 21$

3. Solve each problem any way you choose.

a. $\frac{1}{3} \times 60$ $\qquad$ $\frac{1}{3}$ minute = __________ seconds

b. $\frac{4}{5} \times 60$ $\qquad$ $\frac{4}{5}$ hour = __________ minutes

c. $\frac{7}{10} \times 1000$ $\qquad$ $\frac{7}{10}$ kilogram = __________ grams

d. $\frac{3}{5} \times 100$ $\qquad$ $\frac{3}{5}$ meter = __________ centimeters

1. Convert. Show your work using a tape diagram or an equation.

   a. $\frac{3}{4}$ year = _____ months

   $\frac{3}{4}$ ***year*** $= \frac{3}{4} \times 1$ ***year***

   $= \frac{3}{4} \times 12$ ***months***

   $= \frac{36}{4}$ ***months***

   $= 9$ ***months***

   I can think of $\frac{3}{4}$ year as $\frac{3}{4}$ of 1 year.

   I can rename 1 year as 12 months.

   I can do this in my head: $\frac{3}{4} \times 12 = \frac{3 \times 12}{4} = \frac{36}{4}$.

   b. $\frac{5}{6}$ hour = ______ minutes

   $\frac{5}{6}$ ***hour*** $= \frac{5}{6} \times 1$ ***hour***

   $= \frac{5}{6} \times 60$ ***minutes***

   $= \frac{300}{6}$ ***minutes***

   $= 50$ ***minutes***

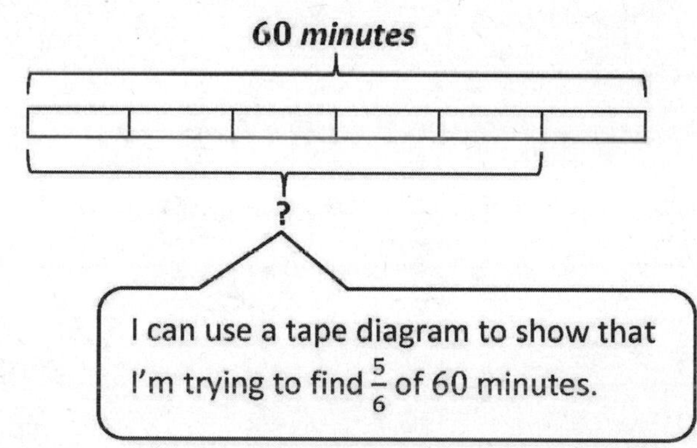

I can use a tape diagram to show that I'm trying to find $\frac{5}{6}$ of 60 minutes.

2. $\frac{2}{3}$ of a yardstick was painted blue. How many feet of the yardstick were painted blue?

$\frac{2}{3}$ ***yard*** = _____ ***feet***

$= \frac{2}{3} \times 1$ ***yard***

$= \frac{2}{3} \times 3$ ***yard***

$= \frac{6}{3}$ ***feet***

$= 2$ ***feet***

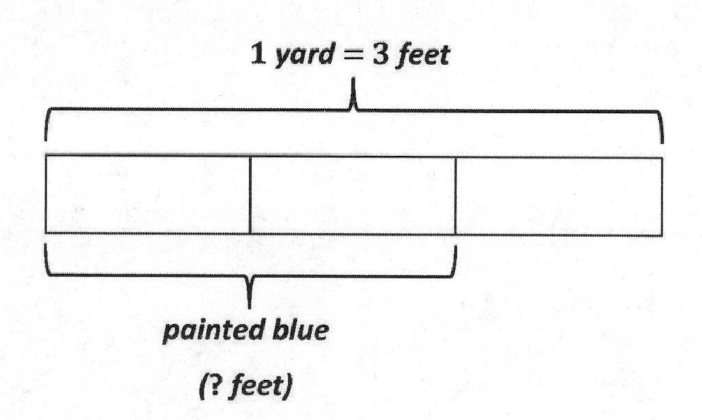

2 feet of the yardstick are painted blue.

Name ______________________________ Date ________________

1. Convert. Show your work using a tape diagram or an equation. The first one is done for you.

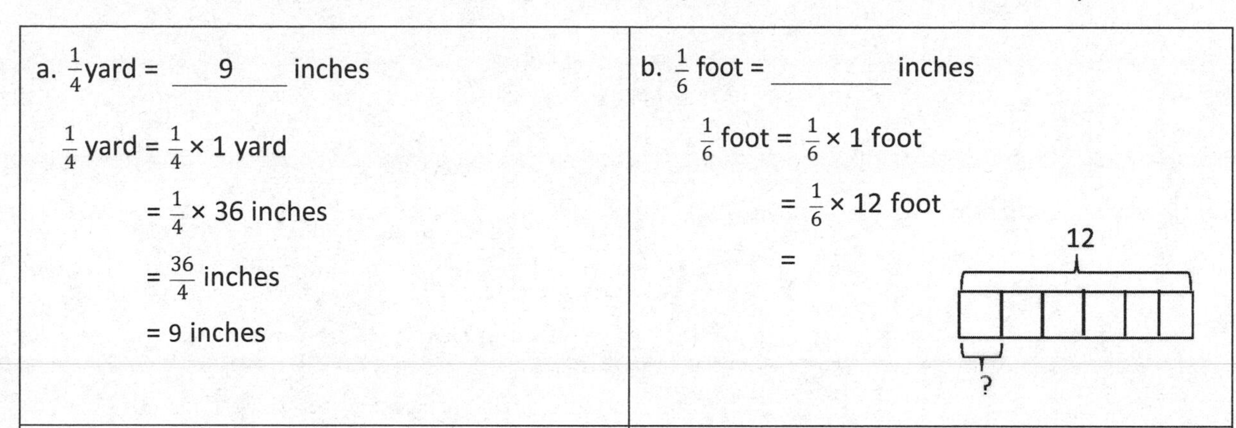

a. $\frac{1}{4}$ yard = ___9___ inches $\frac{1}{4}$ yard = $\frac{1}{4}$ × 1 yard = $\frac{1}{4}$ × 36 inches = $\frac{36}{4}$ inches = 9 inches	b. $\frac{1}{6}$ foot = ________ inches $\frac{1}{6}$ foot = $\frac{1}{6}$ × 1 foot = $\frac{1}{6}$ × 12 foot = 12 ?
c. $\frac{3}{4}$ year = ________ months	d. $\frac{3}{5}$ meter = ________ centimeters
e. $\frac{5}{12}$ hour = ________ minutes	f. $\frac{2}{3}$ yard = ________ inches

2. Michelle measured the length of her forearm. It was $\frac{3}{4}$ of a foot. How long is her forearm in inches?

3. At the market, Ms. Winn bought $\frac{3}{4}$ lb of grapes and $\frac{5}{8}$ lb of cherries.

   a. How many ounces of grapes did Ms. Winn buy?

   b. How many ounces of cherries did Ms. Winn buy?

   c. How many more ounces of grapes than cherries did Ms. Winn buy?

   d. If Mr. Phillips bought $1\frac{3}{4}$ pounds of raspberries, who bought more fruit, Ms. Winn or Mr. Phillips? How many ounces more?

4. A gardener has 10 pounds of soil. He used $\frac{5}{8}$ of the soil for his garden. How many pounds of soil did he use in the garden? How many pounds did he have left?

> *Evaluate* means *solve*, so I need to find the value of the unknown.

1. Write expressions to match the diagrams. Then, evaluate.

a.

> I also could have written $(23 - 8) \times \frac{1}{3}$. Both expressions are correct.

$\frac{1}{3} \times (23 - 8)$

$= \frac{1}{3} \times 15$

$= \frac{15}{3}$

$= 5$

> 23 – 8, or 15, is the whole.

23 – 8

?

> The question mark shows that I'm trying to find 1 third of the whole.

b.

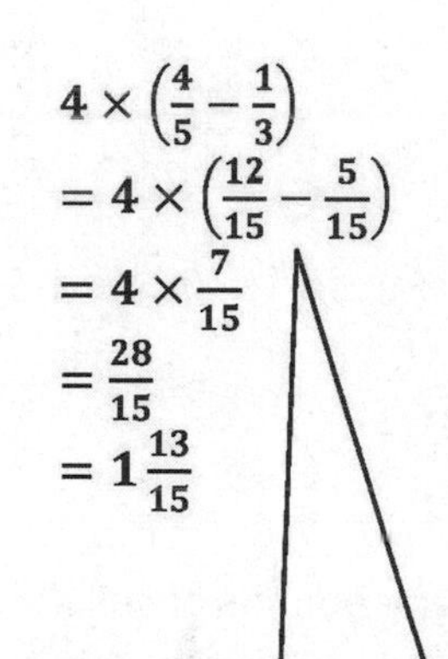

$4 \times \left(\frac{4}{5} - \frac{1}{3}\right)$

$= 4 \times \left(\frac{12}{15} - \frac{5}{15}\right)$

$= 4 \times \frac{7}{15}$

$= \frac{28}{15}$

$= 1\frac{13}{15}$

> In order to subtract, I need to make like units.

> The question mark tells me I need to find the value of the whole.

?

$\frac{4}{5} - \frac{1}{3}$

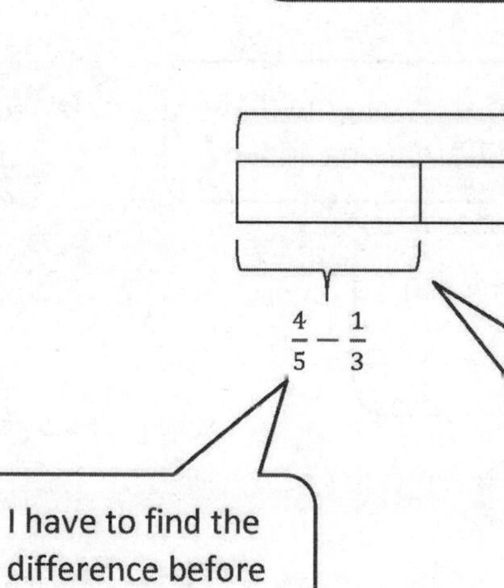

> I have to find the difference before I multiply by 4.

> This 1 unit is equal to $\frac{1}{4}$ of the whole. If I multiply it by 4, I can find the value of the whole.

2. Circle the expression(s) that give the same product as $4 \times \frac{2}{5}$. Explain how you know.

> I can determine which expressions are equivalent to $4 \times \frac{2}{5}$ without evaluating. However, to check my thinking, I can solve. $4 \times \frac{2}{5} = \frac{4 \times 2}{5} = \frac{8}{5} = 1\frac{3}{5}$

a. $5 \div (2 \times 4)$

This expression is equal to $5 \div 8$, not $8 \div 5$.

b. $2 \div 5 \times 4$

$2 \div 5$ is equal to $\frac{2}{5}$. $\frac{2}{5} \times 4 = 4 \times \frac{2}{5}$

c. 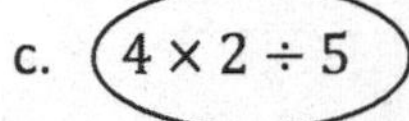
$4 \times 2 \div 5$

This expression is equal to $8 \div 5$, which is $\frac{8}{5}$ or $1\frac{3}{5}$.

d. $4 \times \frac{5}{2}$

This expression does have 4 as one of the factors, but $\frac{5}{2}$ is not equivalent to $\frac{2}{5}$.

3. Write an expression to match, and then evaluate.

a. $\frac{1}{3}$ the sum of 12 and 21

> The word *sum* tells me that 12 and 21 are being added.

> In order to find $\frac{1}{3}$ of the sum, I can multiply by $\frac{1}{3}$ or divide by 3.

$$\frac{1}{3} \times (12 + 21)$$
$$= \frac{1}{3} \times 33$$
$$= \frac{33}{3}$$
$$= 11$$

b. Subtract 5 from $\frac{1}{7}$ of 49.

> I need to be careful with subtraction! Even though the beginning of the expression says to subtract 5, I need to find $\frac{1}{7}$ of 49 first.

$$\frac{1}{7} \times 49 - 5$$
$$= \frac{49}{7} - 5$$
$$= 7 - 5$$
$$= 2$$

4. Use $<$, $>$, or $=$ to make true number sentences without calculating. Explain your thinking.

a. $(17 \times 41) + \frac{5}{4}$ ($<$) $\frac{7}{4} + (17 \times 41)$

Since both expressions show (17×41), I only have to compare the parts being added to this product.

$\frac{5}{4} < \frac{7}{4}$. Therefore, the expression on the left is less than the expression on the right.

In both expressions, one of the factors is $\frac{3}{4}$. I only have to compare the other factors.

I know that $15 + 18 = 33$ and $3 \times 11 = 33$. The second factors are equivalent too.

b. $\frac{3}{4} \times (15 + 18)$ ($=$) $(3 \times 11) \times \frac{3}{4}$

Since both factors are equivalent, these expressions are equal.

Name ______________________________ Date ______________

1. Write expressions to match the diagrams. Then, evaluate.

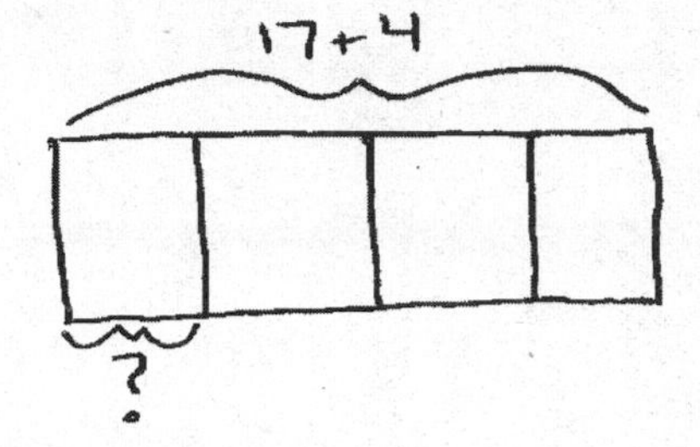

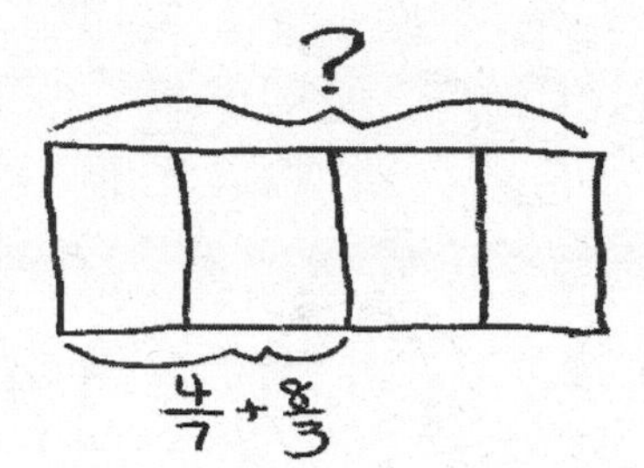

2. Circle the expression(s) that give the same product as $6 \times \frac{3}{8}$. Explain how you know.

$8 \div (3 \times 6)$ $3 \div 8 \times 6$ $(6 \times 3) \div 8$ $(8 \div 6) \times 3$ $6 \times \frac{8}{3}$ $\frac{3}{8} \times 6$

3. Write an expression to match, and then evaluate.

a. $\frac{1}{8}$ the sum of 23 and 17

b. Subtract 4 from $\frac{1}{6}$ of 42.

c. 7 times as much as the sum of $\frac{1}{3}$ and $\frac{4}{5}$

d. $\frac{2}{3}$ of the product of $\frac{3}{8}$ and 16

e. 7 copies of the sum of 8 fifths and 4

f. 15 times as much as 1 fifth of 12

4. Use <, >, or = to make true number sentences without calculating. Explain your thinking.

   a. $\frac{2}{3} \times (9+12)$ 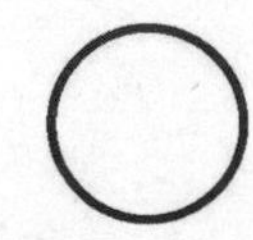 $15 \times \frac{2}{3}$

   b. $\left(3 \times \frac{5}{4}\right) \times \frac{3}{5}$  $\left(3 \times \frac{5}{4}\right) \times \frac{3}{8}$

   c. $6 \times \left(2 + \frac{32}{16}\right)$  $(6 \times 2) + \frac{32}{16}$

5. Fantine bought flour for her bakery each month and recorded the amount in the table to the right. For (a)–(c), write an expression that records the calculation described. Then, solve to find the missing data in the table.

Month	Amount (in pounds)
January	3
February	2
March	$1\frac{1}{4}$
April	
May	$\frac{9}{8}$
June	
July	$1\frac{1}{4}$
August	
September	$\frac{11}{4}$
October	$\frac{3}{4}$

   a. She bought $\frac{3}{4}$ of January's total in August.

   b. She bought $\frac{7}{8}$ as much in April as she did in October and July combined.

c. In June, she bought $\frac{1}{8}$ pound less than three times as much as she bought in May.

d. Display the data from the table in a line plot.

e. How many pounds of flour did Fantine buy from January to October?

Use the RDW (Read, Draw, Write) method to solve.

1. Janice and Adam cooked a 1 lb package of spinach. Janice ate $\frac{1}{2}$ of the spinach, and Adam ate $\frac{1}{4}$ of the spinach. What fraction of the package was left? How many ounces were left?

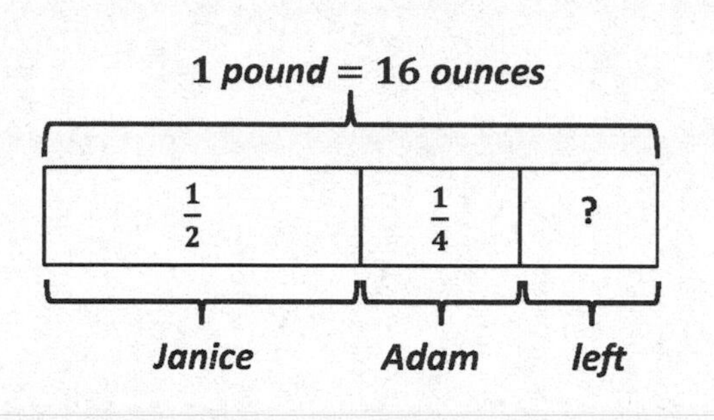

I can add the parts that Janice and Adam ate together to find out what is left over.

$\frac{1}{2} + \frac{1}{4}$

$= \frac{4}{8} + \frac{2}{8}$

$= \frac{6}{8}$

$1 - \frac{6}{8} = \frac{2}{8}$

$\frac{2}{8}$ of the package was left.

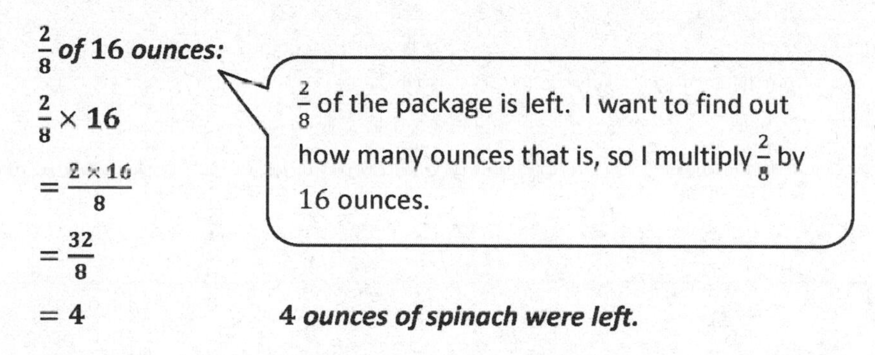

$\frac{2}{8}$ of 16 ounces:

$\frac{2}{8} \times 16$

$= \frac{2 \times 16}{8}$

$= \frac{32}{8}$

$= 4$

$\frac{2}{8}$ of the package is left. I want to find out how many ounces that is, so I multiply $\frac{2}{8}$ by 16 ounces.

4 ounces of spinach were left.

2. Using the tape diagram below, create a story problem about a school. Your story must include a fraction.

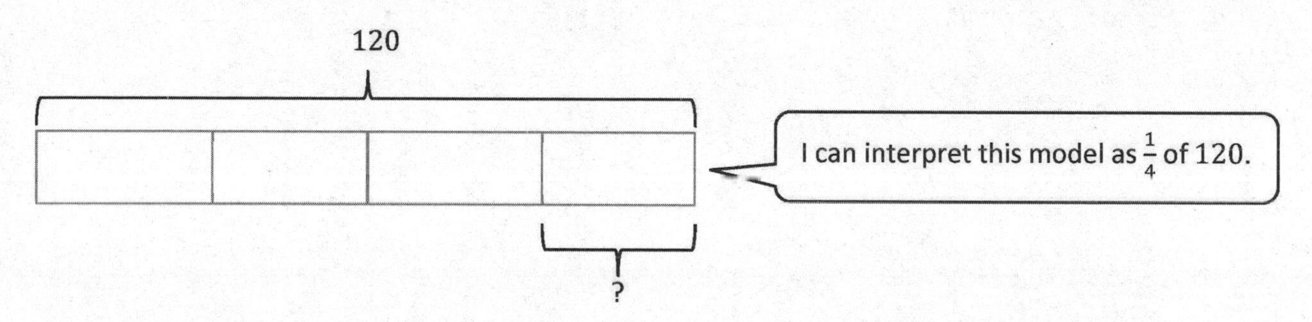

Crestview Elementary School has 120 fifth graders. Three-fourths of them ride the bus to school. The rest of the fifth-grade students walk to school. What fraction of the fifth-grade students walk to school?

Name ______________________________ Date ______________

1. Jenny's mom says she has an hour before it's bedtime. Jenny spends $\frac{1}{3}$ of the hour texting a friend and $\frac{1}{4}$ of the time brushing her teeth and putting on her pajamas. She spends the rest of the time reading her book. How many minutes did Jenny read?

2. A-Plus Auto Body is painting designs on a customer's car. They had 18 pints of blue paint on hand. They used $\frac{1}{2}$ of it for the flames and $\frac{1}{3}$ of it for the sparks. They need $7\frac{3}{4}$ pints of blue paint to paint the next design. How many more pints of blue paint will they need to buy?

3. Giovanna, Frances, and their dad each carried a 10-pound bag of soil into the backyard. After putting soil in the first flower bed, Giovanna's bag was $\frac{5}{8}$ full, Frances's bag was $\frac{2}{5}$ full, and their dad's was $\frac{3}{4}$ full. How many pounds of soil did they put in the first flower bed altogether?

4. Mr. Chan made 252 cookies for the Annual Fifth Grade Class Bake Sale. They sold $\frac{3}{4}$ of them, and $\frac{3}{9}$ of the remaining cookies were given to PTA. members. Mr. Chan allowed the 12 student helpers to divide the cookies that were left equally. How many cookies will each student get?

5. Using the tape diagram below, create a story problem about a farm. Your story must include a fraction.

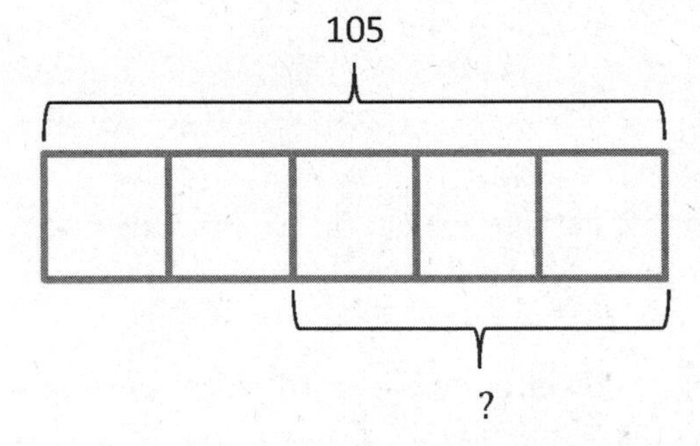

Solve using the RDW (Read, Draw, Write) method.

1. Beth ran her leg of a relay race in $\frac{3}{5}$ the amount of time it took Margaret. Wayne ran his leg of the relay race in $\frac{2}{3}$ the time it took Beth. Margaret finished the race in 30 minutes. How long did it take for Wayne to finish his part of the race?

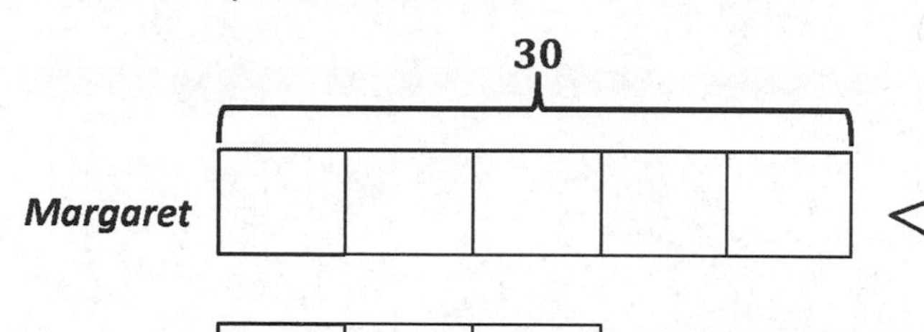

Since Beth's time was $\frac{3}{5}$ of Margaret's, I can partition Margaret's time into 5 equal units. Now I can show that Beth's time is $\frac{3}{5}$ of Margaret's.

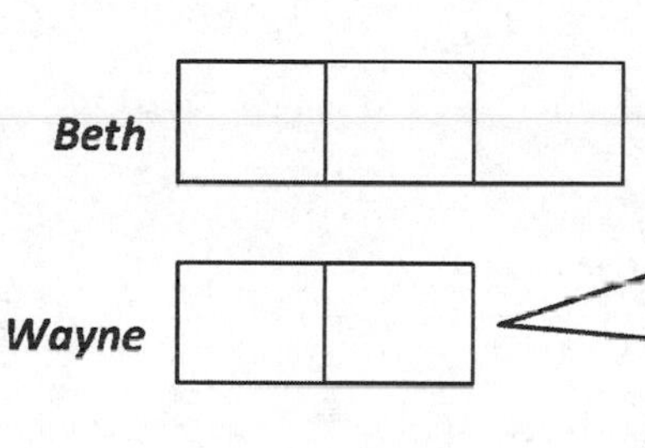

Wayne's time was $\frac{2}{3}$ of Beth's time. 3 units represent Beth's time, so I can show Wayne's time with 2 units. $\frac{2}{3}$ of 3 units is 2 units.

5 units $= 30$

I can use my tape diagram to help me solve. I know that Margaret finished in 30 minutes; therefore, the 5 units representing Margaret's time are equal to 30 minutes.

1 unit $= 30 \div 5 = 6$

I can visualize each unit in the tape diagram being equal to 6 minutes.

2 units $= 2 \times 6 = 12$

Wayne finished the race in 12 minutes.

Wayne's time is equal to 2 units of 6 minutes each, or 12 minutes.

2. Create a story problem about a brother and sister and the money they spend at a deli whose solution is given by the expression $\frac{1}{3} \times (7 + 8)$.

Two siblings went to a deli. The sister had $7.00, and her brother had $8.00. They spent one-third of their combined money. How much money did they spend in the deli?

The parentheses tell me to add first. In my story problem, I wrote that the siblings combined their money.

Name ______________________________ Date ______________

1. Terrence finished a word search in $\frac{3}{4}$ the time it took Frank. Charlotte finished the word search in $\frac{2}{3}$ the time it took Terrence. Frank finished the word search in 32 minutes. How long did it take Charlotte to finish the word search?

2. Ms. Phillips ordered 56 pizzas for a school fundraiser. Of the pizzas ordered, $\frac{2}{7}$ of them were pepperoni, 19 were cheese, and the rest were veggie pizzas. What fraction of the pizzas was veggie?

3. In an auditorium, $\frac{1}{6}$ of the students are fifth graders, $\frac{1}{3}$ are fourth graders, and $\frac{1}{4}$ of the remaining students are second graders. If there are 96 students in the auditorium, how many second graders are there?

4. At a track meet, Jacob and Daniel compete in the 220 m hurdles. Daniel finishes in $\frac{3}{4}$ of a minute. Jacob finishes with $\frac{5}{12}$ of a minute remaining. Who ran the race in the faster time?

Bonus: Express the difference in their times as a fraction of a minute.

5. Create and solve a story problem about a runner who is training for a race. Include at least one fraction in your story.

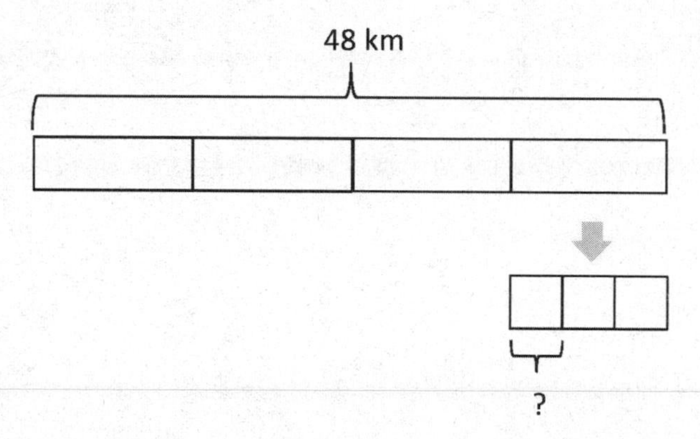

6. Create and solve a story problem about two friends and their weekly allowance whose solution is given by the expression $\frac{1}{5} \times (12 + 8)$.

1. Solve. Draw a rectangular fraction model to show your thinking.

   a. Half of $\frac{1}{4}$ pan of brownies

   Half of $\frac{1}{4} = \frac{1}{8}$

   $\frac{1}{2} \times \frac{1}{4} = \frac{1}{8}$

The problem tells me I have $\frac{1}{4}$ pan of brownies. I can draw a whole pan. Then, I can shade and label $\frac{1}{4}$ of the pan.

Seeing the word *of* reminds me of third grade when I learned that 2 × 3 meant 2 groups *of* 3.

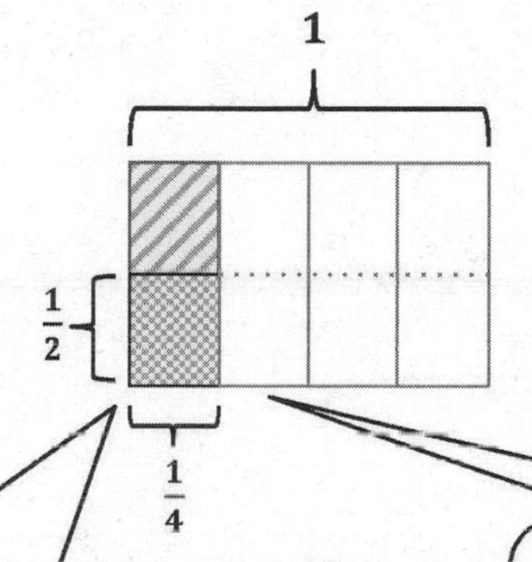

Since I want to model 1 half of the fourth of a pan, I can partition the fourth into 2 equal parts, or halves. I can shade $\frac{1}{2}$ of the $\frac{1}{4}$.

My model shows me that $\frac{1}{2}$ of $\frac{1}{4}$ is equal to $\frac{1}{8}$ of the pan of brownies.

   b. $\frac{1}{4} \times \frac{1}{4}$

   $\frac{1}{4} \times \frac{1}{4} = \frac{1}{16}$

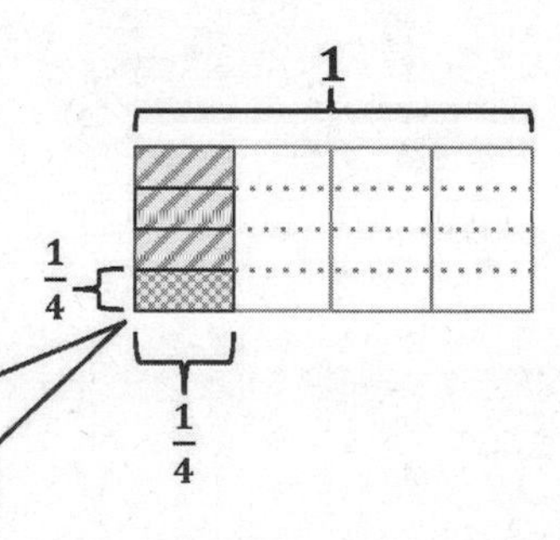

The part that is double-shaded shows $\frac{1}{4}$ of $\frac{1}{4}$.

$\frac{1}{4}$ of $\frac{1}{4} = \frac{1}{16}$

2. The Guerra family uses $\frac{3}{4}$ of their backyard for a pool. $\frac{1}{3}$ of the remaining yard is used for a vegetable garden. The rest of the yard is grass. What fraction of the entire backyard is for the vegetable garden? Draw a picture to support your answer.

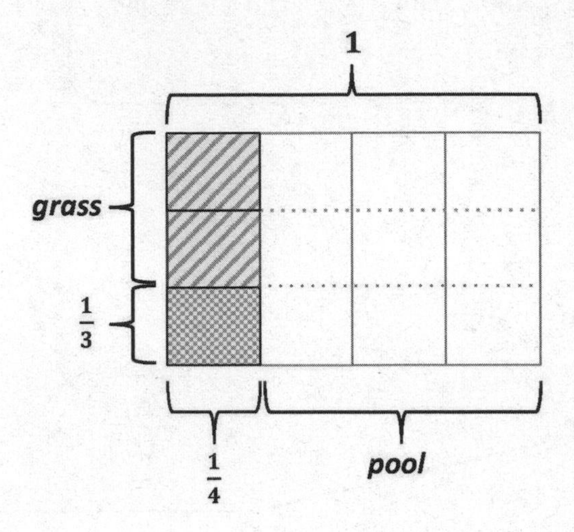

Since $\frac{3}{4}$ of the backyard is a pool, that means $\frac{1}{4}$ of the backyard is *not* a pool.

$$\frac{1}{3} \times \frac{1}{4} = \frac{1}{12}$$

$\frac{1}{12}$ of the backyard is a vegetable garden.

Name ______________________________ Date ________________

1. Solve. Draw a rectangular fraction model to show your thinking.

a. Half of $\frac{1}{2}$ cake = _____ cake.

b. One-third of $\frac{1}{2}$ cake = _____ cake.

c. $\frac{1}{4}$ of $\frac{1}{2}$

d. $\frac{1}{2} \times \frac{1}{5}$

e. $\frac{1}{3} \times \frac{1}{3}$

f. $\frac{1}{4} \times \frac{1}{3}$

2. Noah mows $\frac{1}{2}$ of his property and leaves the rest wild. He decides to use $\frac{1}{5}$ of the wild area for a vegetable garden. What fraction of the property is used for the garden? Draw a picture to support your answer.

3. Fawn plants $\frac{2}{3}$ of the garden with vegetables. Her son plants the remainder of the garden. He decides to use $\frac{1}{2}$ of his space to plant flowers, and in the rest, he plants herbs. What fraction of the entire garden is planted in flowers? Draw a picture to support your answer.

4. Diego eats $\frac{1}{5}$ of a loaf of bread each day. On Tuesday, Diego eats $\frac{1}{4}$ of the day's portion before lunch. What fraction of the whole loaf does Diego eat before lunch on Tuesday? Draw a rectangular fraction model to support your thinking.

1. Solve. Draw a rectangular fraction model to explain your thinking.

a. $\frac{1}{3}$ of $\frac{3}{5}$ = $\frac{1}{3}$ of **3** fifths = **1** fifth

$\frac{1}{3}$ of 3 is 1.

$\frac{1}{3}$ of 3 bananas is 1 banana.

$\frac{1}{3}$ of 3 fifths is 1 fifth.

$$\frac{1}{3} \times \frac{3}{5} = \frac{3}{15} = \frac{1}{5}$$

I can model $\frac{3}{5}$ by partitioning vertically first. Then to show $\frac{1}{3}$ of $\frac{3}{5}$, I can partition with horizontal lines.

b. $\frac{1}{2} \times \frac{3}{4}$

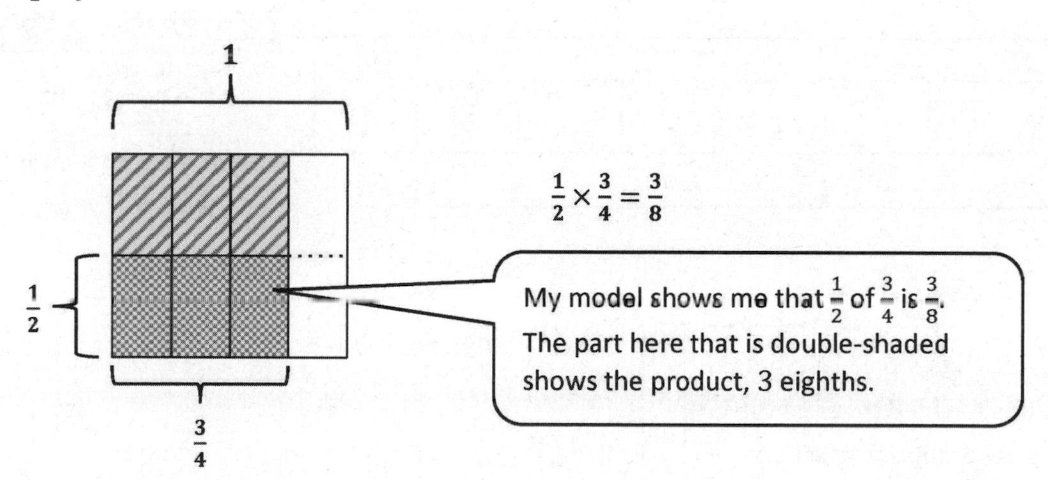

2. Kenny collects coins. $\frac{3}{5}$ of his collection is dimes. $\frac{1}{2}$ of the remaining coins are quarters. What fraction of Kenny's whole collection is quarters? Support your answer with a model.

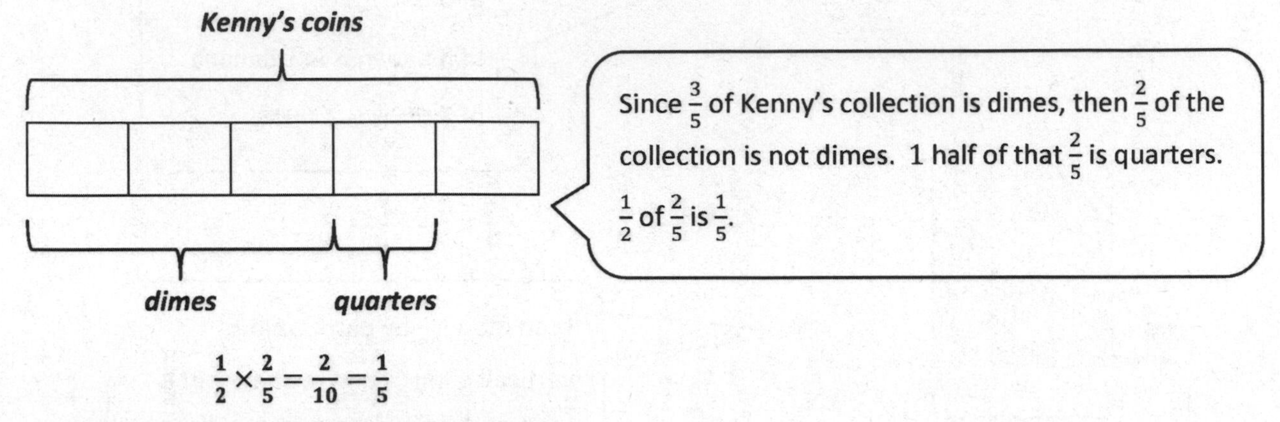

$$\frac{1}{2} \times \frac{2}{5} = \frac{2}{10} = \frac{1}{5}$$

One fifth of Kenny's coin collection is quarters.

3. In Jan's class, $\frac{3}{8}$ of the students take the bus to school. $\frac{4}{5}$ of the non-bus riders walk to school. One half of the remaining students ride their bikes to school.

   a. What fraction of all the students walk to school?

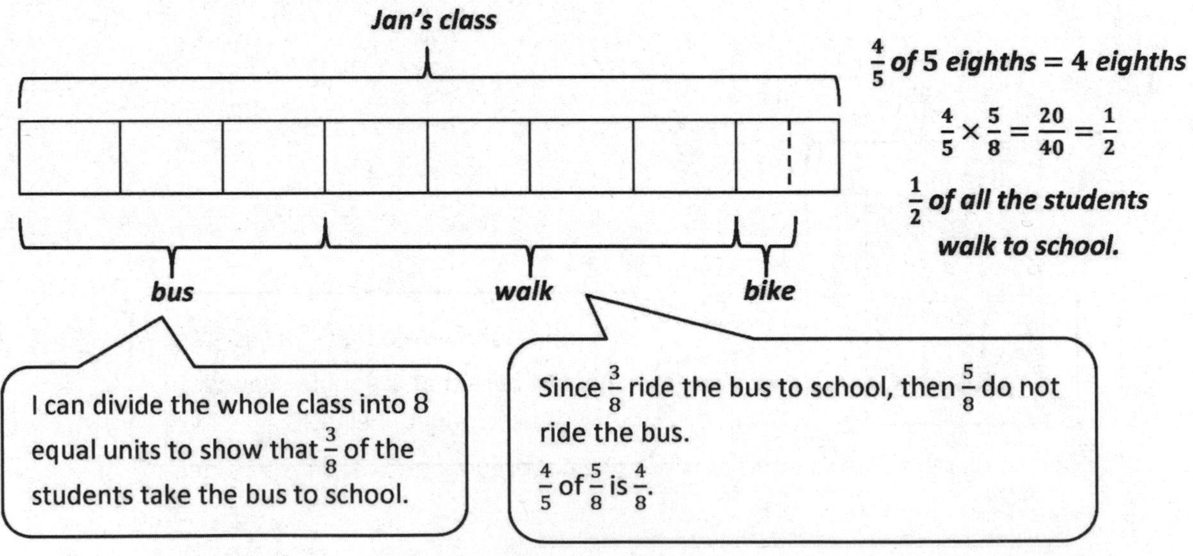

   b. What fraction of all the students ride their bikes to school?

$\frac{1}{2}$ ***of*** $\frac{1}{8} = \frac{1}{16}$

$\frac{1}{16}$ ***of all the students bike to school.***

After labeling the units that represent the students that walk or bus to school, there was only 1 unit, or $\frac{1}{8}$ of the class, remaining. Half of those students bike to school.

Name ____________________ Date ____________

1. Solve. Draw a rectangular fraction model to explain your thinking.

a. $\frac{1}{2}$ of $\frac{2}{3}$ = $\frac{1}{2}$ of ____ third(s) = ____ third(s)

b. $\frac{1}{2}$ of $\frac{4}{3}$ = $\frac{1}{2}$ of ____ third(s) = ____ third(s)

c. $\frac{1}{3}$ of $\frac{3}{5}$ =

d. $\frac{1}{2}$ of $\frac{6}{8}$ =

e. $\frac{1}{3} \times \frac{4}{5}$ =

f. $\frac{4}{5} \times \frac{1}{3}$ =

2. Sarah has a photography blog. $\frac{3}{7}$ of her photos are of nature. $\frac{1}{4}$ of the rest are of her friends. What fraction of all of Sarah's photos is of her friends? Support your answer with a model.

3. At Laurita's Bakery, $\frac{3}{5}$ of the baked goods are pies, and the rest are cakes. $\frac{1}{3}$ of the pies are coconut. $\frac{1}{6}$ of the cakes are angel food.

   a. What fraction of all of the baked goods at Laurita's Bakery are coconut pies?

   b. What fraction of all of the baked goods at Laurita's Bakery are angel food cakes?

4. Grandpa Mick opened a pint of ice cream. He gave his youngest grandchild $\frac{1}{5}$ of the ice cream and his middle grandchild $\frac{1}{4}$ of the remaining ice cream. Then, he gave his oldest grandchild $\frac{1}{3}$ of the ice cream that was left after serving the others.

   a. Who got the most ice cream? How do you know? Draw a picture to support your reasoning.

   b. What fraction of the pint of ice cream will be left if Grandpa Mick serves himself the same amount as the second grandchild?

1. Solve. Draw a rectangular fraction model to explain your thinking. Then, write a multiplication sentence.

   $\frac{2}{5}$ of $\frac{2}{3}$

   $\frac{2}{5} \times \frac{2}{3} = \frac{4}{15}$

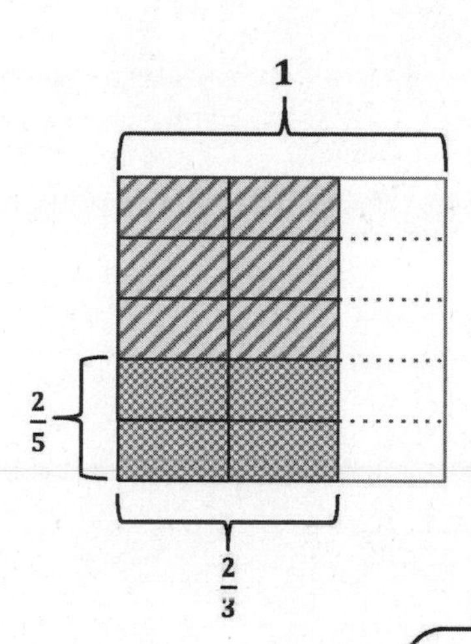

2. Multiply.

   a. $\frac{3}{8} \times \frac{2}{5}$

   The 2 in the numerator and the 8 in the denominator have a common factor of 2.

   $2 \div 2 = 1$ and $8 \div 2 = 4$

   $$\frac{3}{8} \times \frac{2}{5} = \frac{3 \times \overset{1}{\cancel{2}}}{\underset{4}{\cancel{8}} \times 5} = \frac{3}{20}$$

   Now the numerator is 3×1, and the denominator is 4×5.

   b. $\frac{2}{5} \times \frac{10}{12}$

   I was able to rename this fraction twice before multiplying. 5 and 10 have a common factor of 5.

   $$\frac{2}{5} \times \frac{10}{12} = \frac{\overset{1}{\cancel{2}} \times \overset{2}{\cancel{10}}}{\underset{1}{\cancel{5}} \times \underset{6}{\cancel{12}}} = \frac{2}{6}$$

   And 2 and 12 have a common factor of 2.

   Now the numerator is 1×2, and the denominator is 1×6.

Name ______________________________ Date ______________

1. Solve. Draw a rectangular fraction model to explain your thinking. Then, write a multiplication sentence.

a. $\frac{2}{3}$ of $\frac{3}{4}$ =

b. $\frac{2}{5}$ of $\frac{3}{4}$ =

c. $\frac{2}{5}$ of $\frac{4}{5}$ =

d. $\frac{4}{5}$ of $\frac{3}{4}$ =

2. Multiply. Draw a rectangular fraction model if it helps you.

a. $\frac{5}{6} \times \frac{3}{10}$

b. $\frac{3}{4} \times \frac{4}{5}$

c. $\frac{5}{6} \times \frac{5}{8}$

d. $\frac{3}{4} \times \frac{5}{12}$

e. $\frac{8}{9} \times \frac{2}{3}$

f. $\frac{3}{7} \times \frac{2}{9}$

3. Every morning, Halle goes to school with a 1-liter bottle of water. She drinks $\frac{1}{4}$ of the bottle before school starts and $\frac{2}{3}$ of the rest before lunch.

   a. What fraction of the bottle does Halle drink after school starts but before lunch?

   b. How many milliliters are left in the bottle at lunch?

4. Moussa delivered $\frac{3}{8}$ of the newspapers on his route in the first hour and $\frac{4}{5}$ of the rest in the second hour. What fraction of the newspapers did Moussa deliver in the second hour?

5. Rose bought some spinach. She used $\frac{3}{5}$ of the spinach on a pan of spinach pie for a party and $\frac{3}{4}$ of the remaining spinach for a pan for her family. She used the rest of the spinach to make a salad.

   a. What fraction of the spinach did she use to make the salad?

   b. If Rose used 3 pounds of spinach to make the pan of spinach pie for the party, how many pounds of spinach did Rose use to make the salad?

Solve and show your thinking with a tape diagram.

1. Heidi had 6 pounds of tomatoes from her garden. She used $\frac{3}{4}$ of all the tomatoes to make sauce and gave $\frac{2}{3}$ of the rest of the tomatoes to her neighbor. How many ounces of tomatoes did Heidi give to her neighbor?

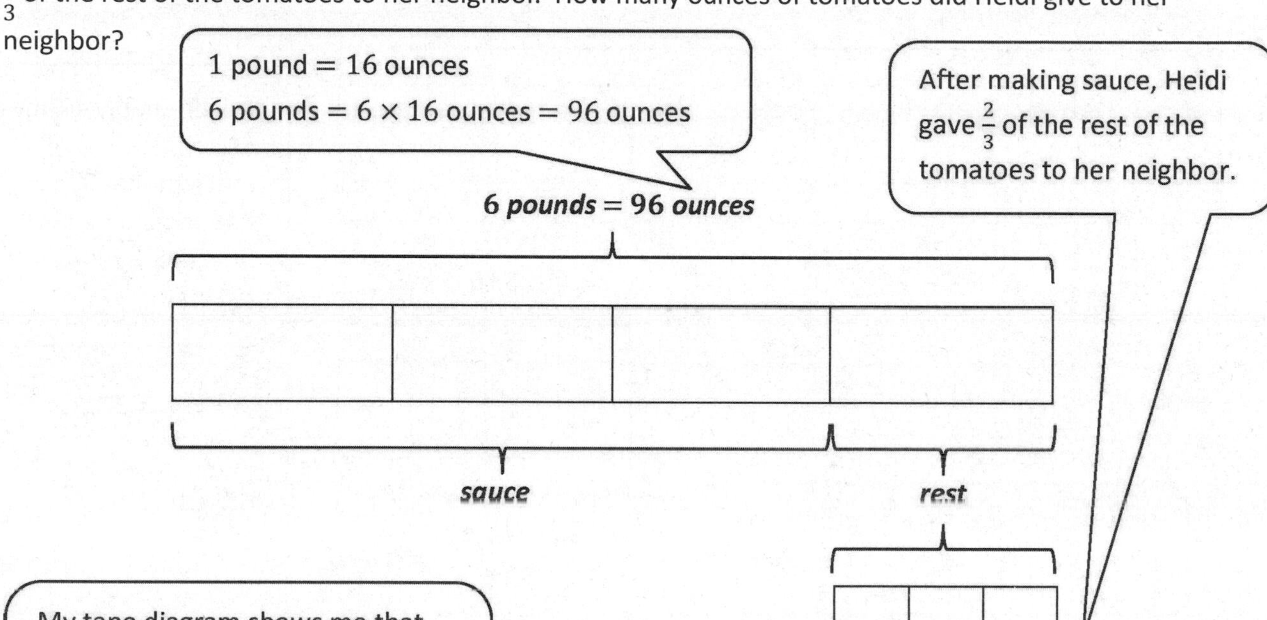

My tape diagram shows me that the total value of the 4 units is 96 ounces. I can divide to find the value of 1 unit, or $\frac{1}{4}$ of 96.

neighbor
(? ounces)

4 units = 96

1 unit = 96 ÷ 4 = 24

Now I know that Heidi had 24 ounces of tomatoes left after making sauce.

When I look at my model, I can think of this another way. Heidi gave $\frac{2}{3}$ of $\frac{1}{4}$ to her neighbor.

$\frac{2}{3} \times \frac{1}{4} = \frac{2}{12} = \frac{1}{6}$

Heidi gave $\frac{1}{6}$ of all the tomatoes to her neighbor.

$\frac{1}{6}$ of 96 = 16

$\frac{2}{3}$ *of* $24 = \frac{2 \times 24}{3} = \frac{48}{3} = 16$

I can find $\frac{2}{3}$ of 24 and know how many ounces Heidi gave to her neighbor.

Heidi gave her neighbor 16 ounces of tomatoes.

2. Tracey spent $\frac{2}{3}$ of her money on movie tickets and $\frac{3}{4}$ of the remaining money on popcorn and water. If she had $4 left over, how much money did she have at first?

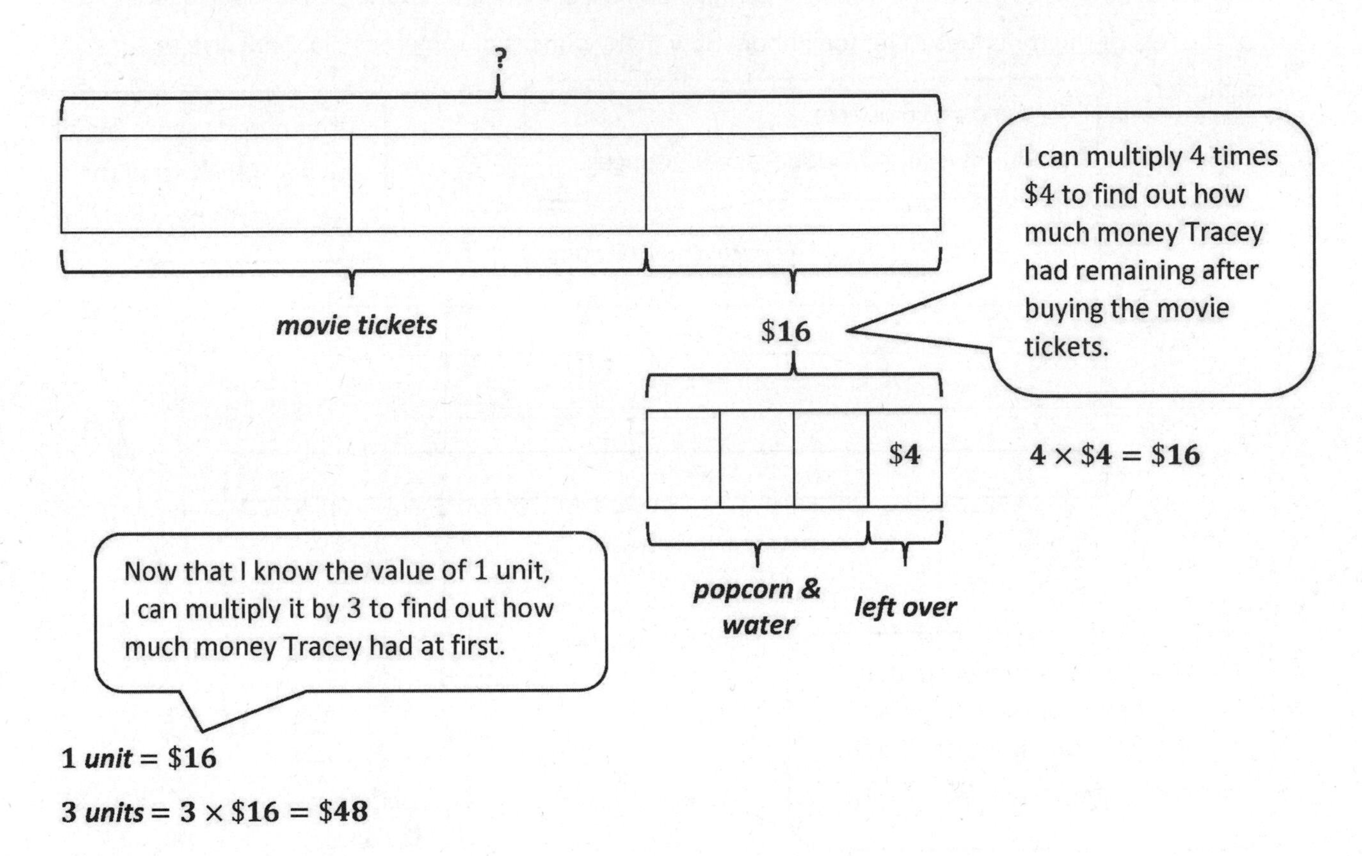

1 unit = $16

3 units = 3 × $16 = $48

Tracey had $48 *at first.*

Name ______________________________ Date ______________

Solve and show your thinking with a tape diagram.

1. Anthony bought an 8-foot board. He cut off $\frac{3}{4}$ of the board to build a shelf and gave $\frac{1}{3}$ of the rest to his brother for an art project. How many inches long was the piece Anthony gave to his brother?

2. Riverside Elementary School is holding a school-wide election to choose a school color. Five-eighths of the votes were for blue, $\frac{5}{9}$ of the remaining votes were for green, and the remaining 48 votes were for red.

   a. How many votes were for blue?

   b. How many votes were for green?

c. If every student got one vote, but there were 25 students absent on the day of the vote, how many students are there at Riverside Elementary School?

d. Seven-tenths of the votes for blue were made by girls. Did girls who voted for blue make up more than or less than half of all votes? Support your reasoning with a picture.

e. How many girls voted for blue?

1. Multiply and model. Rewrite each expression as a multiplication sentence with decimal factors.

a. $\frac{3}{10} \times \frac{2}{10}$

$= \frac{3 \times 2}{10 \times 10}$

$= \frac{6}{100}$

When multiplying fractions, I multiply the two numerators, 3×2, and the two denominators, 10×10, to get $\frac{6}{100}$.

Since the whole grid represents 1, each square represents $\frac{1}{100}$. 10 squares is equal to $\frac{1}{10}$.

I shade in $\frac{2}{10}$ (20 squares vertically).

I shade in $\frac{3}{10}$ of $\frac{2}{10}$ (6 squares).

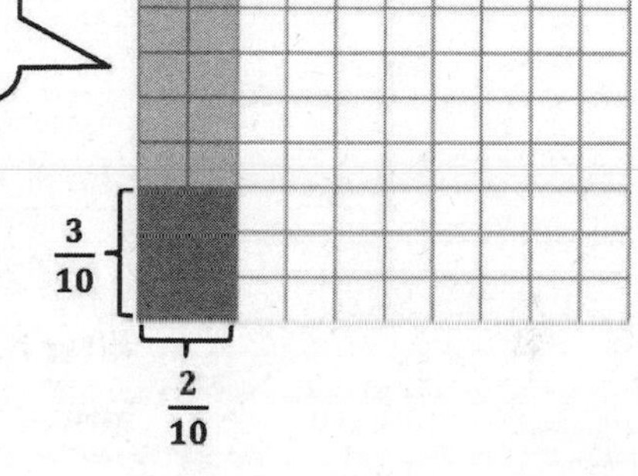

I label each whole grid as 1, and each square represents $\frac{1}{100}$.

b. $\frac{3}{10} \times 1.2$

$= \frac{3}{10} \times \frac{12}{10}$

$= \frac{3 \times 12}{10 \times 10}$

$= \frac{36}{100}$

I shade in 1 and $\frac{2}{10}$ (120 squares vertically).

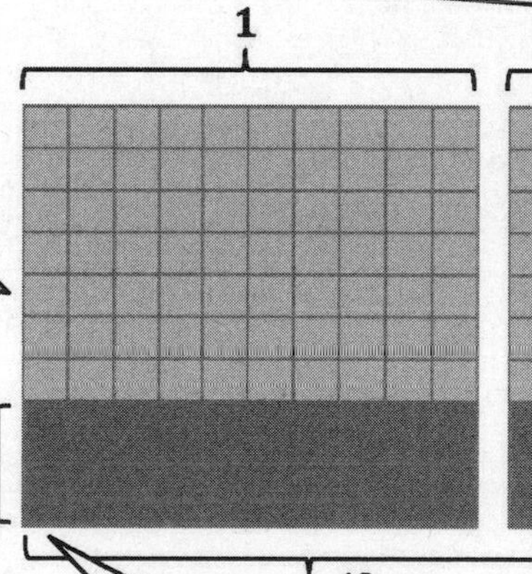

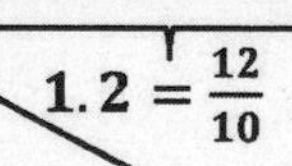

I rename 1.2 as a fraction greater than one, $\frac{12}{10}$, and then multiply to get $\frac{36}{100}$.

I shade in $\frac{3}{10}$ of $\frac{12}{10}$ (36 squares).

2. Multiply.

   a. 2×0.6

   $= 2 \times \frac{6}{10}$

   $= \frac{2 \times 6}{10}$

   $= \frac{12}{10}$

   $= 1.2$

   I rewrite the decimal as a fraction and then multiply the two numerators and the two denominators to get $\frac{12}{10}$. Lastly, I write it as a mixed number if possible.

   b. 0.2×0.6

   $= \frac{2}{10} \times \frac{6}{10}$

   $= \frac{2 \times 6}{10 \times 10}$

   $= \frac{12}{100}$

   $= 0.12$

   0.2 is 2 tenths, or $\frac{2}{10}$. After multiplying, the answer is $\frac{12}{100}$, or 0.12.

   c. 0.02×0.6

   $= \frac{2}{100} \times \frac{6}{10}$

   $= \frac{2 \times 6}{100 \times 10}$

   $= \frac{12}{1,000}$

   $= 0.012$

   0.02 is 2 hundredths, or $\frac{2}{100}$. After multiplying, the answer is $\frac{12}{1,000}$ or 0.012.

3. Sydney makes 1.2 liters of orange juice. If she pours 4 tenths of the orange juice in the glass, how many liters of orange juice are in the glass?

   $\frac{4}{10}$ *of* 1.2 L

   $\frac{4}{10} \times 1.2$

   $= \frac{4}{10} \times \frac{12}{10}$

   $= \frac{4 \times 12}{10 \times 10}$

   $= \frac{48}{100}$

   $= 0.48$

   To find 4 tenths of 1.2 liters, I multiply $\frac{4}{10}$ times $\frac{12}{10}$ to get $\frac{48}{100}$, or 0.48.

   There are 0.48 L ***of orange juice in the glass.***

Name ______________________________ Date ______________

1. Multiply and model. Rewrite each expression as a number sentence with decimal factors. The first one is done for you.

a. $\frac{1}{10} \times \frac{1}{10}$

$= \frac{1 \times 1}{10 \times 10}$

$= \frac{1}{100}$

$0.1 \times 0.1 = 0.01$

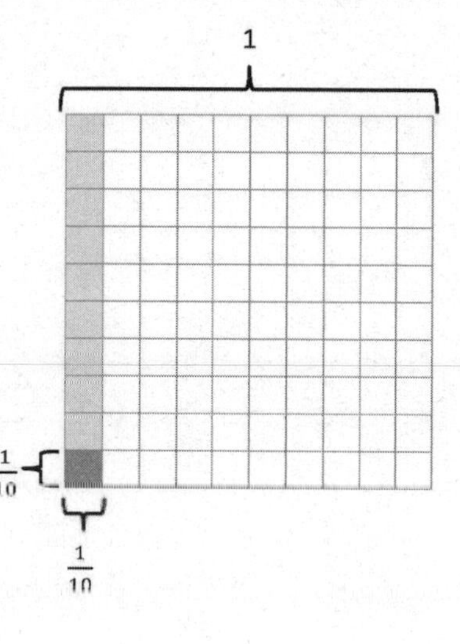

b. $\frac{6}{10} \times \frac{2}{10}$

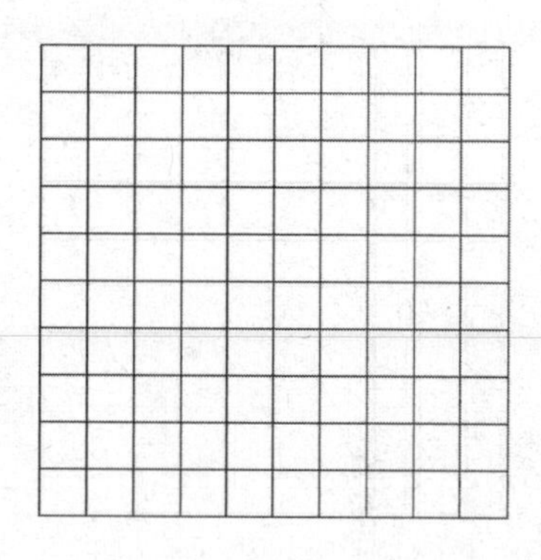

c. $\frac{1}{10} \times 1.6$

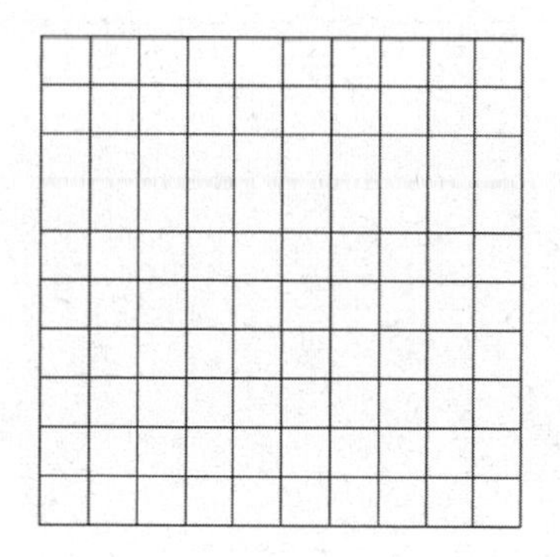

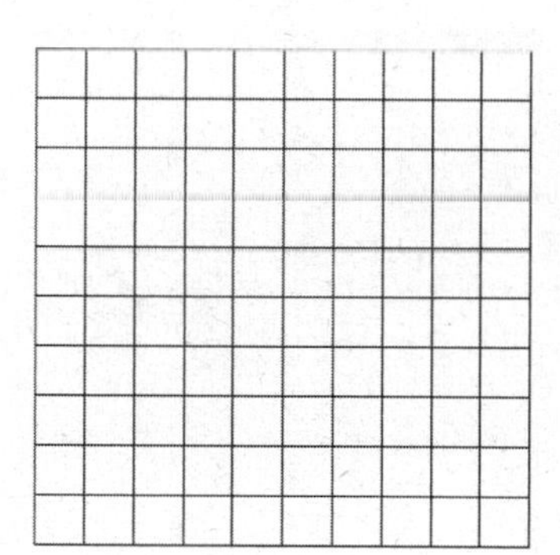

d. $\frac{6}{10} \times 1.9$

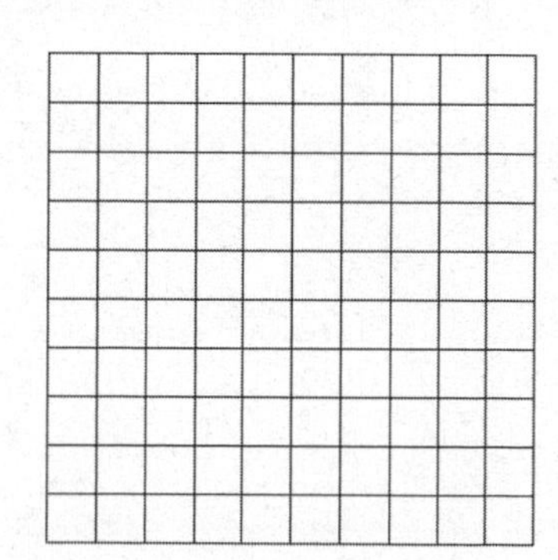

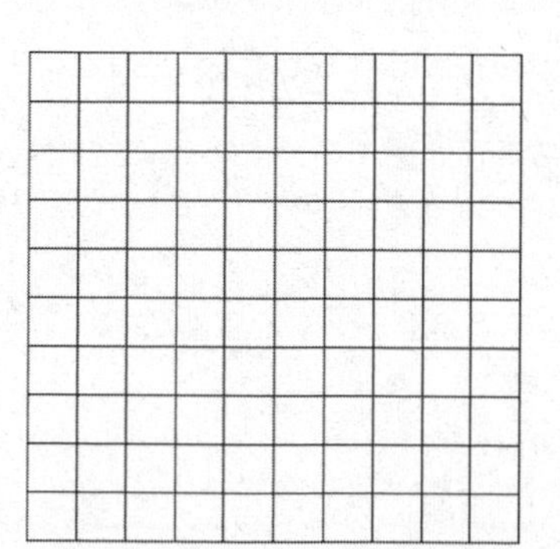

2. Multiply. The first few are started for you.

a. $4 \times 0.6 =$ ______

$= 4 \times \frac{6}{10}$

$= \frac{4 \times 6}{10}$

$= \frac{24}{10}$

$= 2.4$

b. $0.4 \times 0.6 =$ ______

$= \frac{4}{10} \times \frac{6}{10}$

$= \frac{4 \times 6}{10 \times 10}$

$=$

c. $0.04 \times 0.6 =$ ______

$= \frac{4}{100} \times \frac{6}{10}$

$= \frac{\_\_ \times \_\_}{100 \times 10}$

$=$

d. $7 \times 0.3 =$ ______

e. $0.7 \times 0.3 =$ ______

f. $0.07 \times 0.3 =$ ______

g. $1.3 \times 5 =$ ______

h. $1.3 \times 0.5 =$ ______

i. $0.13 \times 0.5 =$ ______

3. Jennifer makes 1.7 liters of lemonade. If she pours 3 tenths of the lemonade in the glass, how many liters of lemonade are in the glass?

4. Cassius walked 6 tenths of a 3.6-mile trail.

a. How many miles did Cassius have left to hike?

b. Cameron was 1.3 miles ahead of Cassius. How many miles did Cameron hike already?

1. Multiply using both fraction form and unit form.

a. $2.3 \times 1.6 = \frac{23}{10} \times \frac{16}{10}$

$= \frac{23 \times 16}{10 \times 10}$

$= \frac{368}{100}$

$= 3.68$

```
      2  3  tenths
×     1  6  tenths
   ------------
   1  3  8
+  2  3  0
   ------------
   3  6  8  hundredths
```

I write the decimals (2.3 and 1.6) in unit form (23 tenths and 16 tenths).

I express the decimals (2.3 and 1.6) as fractions ($\frac{23}{10}$ and $\frac{16}{10}$), and then I multiply to get $\frac{368}{100}$, or 3.68.

I multiply the 2 factors as if they are whole numbers to get 368. The product's unit is hundredths because a tenth times a tenth is equal to a hundredth.

b. $2.38 \times 1.8 = \frac{238}{100} \times \frac{18}{10}$

$= \frac{238 \times 18}{100 \times 10}$

$= \frac{4{,}284}{1{,}000}$

$= 4.284$

```
       2  3  8  hundredths
×         1  8  tenths
   ----------------
    1  9  0  4
+   2  3  8  0
   ----------------
   4,  2  8  4  thousandths
```

I express the decimals (2.38 and 1.8) in unit form (238 hundredths and 18 tenths).

A hundredth times a tenth is a thousandth.

2. A flower garden measures 2.75 meters by 4.2 meters.

   a. Find the area of the flower garden.

$\mathbf{2.75\ m \times 4.2\ m = 11.55\ m^2}$

The area of the flower garden is* 11.55 *square meters.

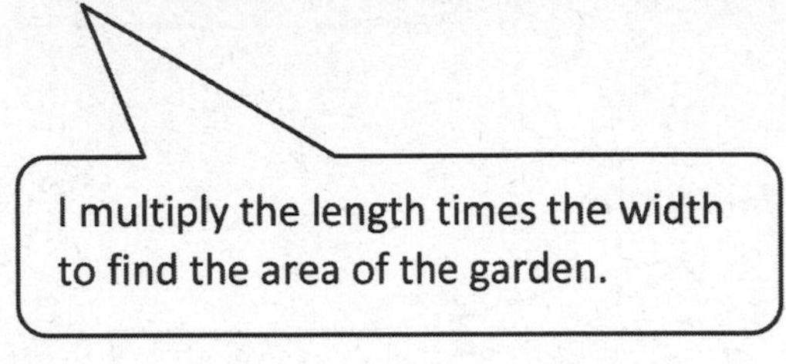

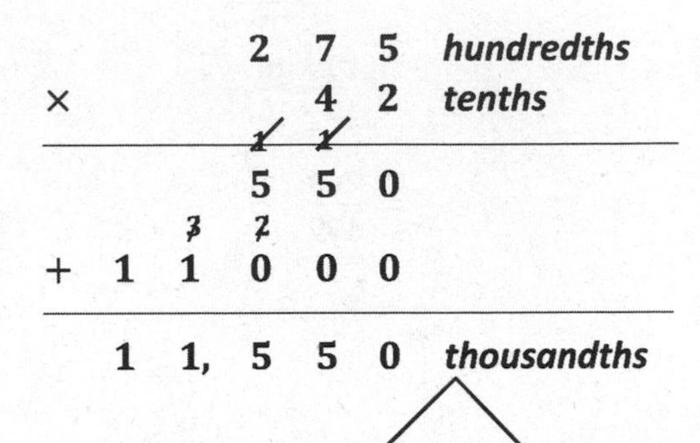

A hundredth times a tenth is a thousandth.

$\frac{1}{100} \times \frac{1}{10} = \frac{1 \times 1}{100 \times 10} = \frac{1}{1{,}000}$

   b. The area of the vegetable garden is one and a half times that of the flower garden. Find the total area of the flower garden and the vegetable garden.

$\mathbf{11.55\ m^2 \times 1.5 = 17.325\ m^2}$

$\mathbf{11.55\ m^2 + 17.325\ m^2 = 28.875\ m^2}$

```
      1 1 5 5   hundredths
  ×       1 5   tenths
  -------------
      5 7 7 5
  + 1 1 5 5 0
  -------------
    1 7, 3 2 5  thousandths
```

I find the area of the vegetable garden by multiplying the flower garden's area by 1.5, or 15 tenths.

```
    1 1. 5 5 0
  + 1 7. 3 2 5
  ------------
    2 8. 8 7 5
```

I add the 2 areas together to find the total area.

***The total area of the flower garden and the vegetable garden is* 28.875 m².**

Name ______________________________ Date ____________

1. Multiply using fraction form and unit form. Check your answer by counting the decimal places. The first one is done for you.

a. $3.3 \times 1.6 = \frac{33}{10} \times \frac{16}{10}$

$= \frac{33 \times 16}{100}$

$= \frac{528}{100}$

$= 5.28$

```
    3 3 tenths
×   1 6 tenths
    1 9 8
+   3 3 0
    5 2 8 hundredths
```

b. $3.3 \times 0.8 =$

```
  3 3 tenths
×   8 tenths
```

c. $4.4 \times 3.2 =$

d. $2.2 \times 1.6 =$

2. Multiply using fraction form and unit form. The first one is partially done for you.

a. $3.36 \times 1.4 = \frac{336}{100} \times \frac{14}{10}$

$= \frac{336 \times 14}{1{,}000}$

$= \frac{4{,}704}{1{,}000}$

$= 4.704$

```
  3 3 6 hundredths
×     1 4 tenths
```

b. $3.35 \times 0.7 =$

```
  3 3 5 hundredths
×       7 tenths
```

c. $4.04 \times 3.2 =$

d. $4.4 \times 0.16 =$

3. Solve using the standard algorithm. Show your thinking about the units of your product. The first one is done for you.

a. $3.2 \times 0.6 = 1.92$ $\frac{32}{10} \times \frac{6}{10} = \frac{32 \times 6}{100}$

```
  3 2 tenths
×   6 tenths
1 9 2 hundredths
```

b. $2.3 \times 2.1 =$ ________

```
  2 3 tenths
× 2 1 tenths
```

c. $7.41 \times 3.4 =$ ________

d. $6.50 \times 4.5 =$ ________

4. Erik buys 2.5 pounds of cashews. If each pound of cashews costs $7.70, how much will he pay for the cashews?

5. A swimming pool at a park measures 9.75 meters by 7.2 meters.

a. Find the area of the swimming pool.

b. The area of the playground is one and a half times that of the swimming pool. Find the total area of the swimming pool and the playground.

1. Convert. Express your answer as a mixed number, if possible.

   a. 9 in = _____ ft

   I know that 1 foot = 12 inches and 1 inch = $\frac{1}{12}$ foot.

   $$\mathbf{9\text{ in} = 9 \times 1\text{ in}}$$
   $$\mathbf{= 9 \times \frac{1}{12}\text{ ft}}$$
   $$\mathbf{= \frac{9}{12}\text{ ft}}$$
   $$\mathbf{= \frac{3}{4}\text{ ft}}$$

   9 inches is equal to 9 times 1 inch. I can rename 1 inch as $\frac{1}{12}$ foot and then multiply.

   b. 20 oz = _____ lb

   I know that 1 pound = 16 ounces and 1 ounce = $\frac{1}{16}$ pound.

   $$\mathbf{20\text{ oz} = 20 \times 1\text{ oz}}$$
   $$\mathbf{= 20 \times \frac{1}{16}\text{ lb}}$$
   $$\mathbf{= \frac{20}{16}\text{ lb}}$$
   $$\mathbf{= 1\frac{4}{16}\text{ lb}}$$
   $$\mathbf{= 1\frac{1}{4}\text{ lb}}$$

   20 ounces is equal to 20 times 1 ounce. I can rename 1 ounce as $\frac{1}{16}$ pound and then multiply.

2. Jack buys 14 ounces of peanuts.
   What fraction of a pound of peanuts did Jack buy?

   14 oz = _____ lb

   1 pound = 16 ounces, and 1 ounce = $\frac{1}{16}$ pound.

   $$\mathbf{14\text{ oz} = 14 \times 1\text{ oz}}$$
   $$\mathbf{= 14 \times \frac{1}{16}\text{ lb}}$$
   $$\mathbf{= \frac{14}{16}\text{ lb}}$$
   $$\mathbf{= \frac{7}{8}\text{ lb}}$$

   Jack bought $\frac{7}{8}$ pound of peanuts.

Name ____________________ Date __________

1. Convert. Express your answer as a mixed number, if possible.

a. 2 ft = $\frac{2}{3}$ yd 2 ft = 2 × 1 ft $= 2 \times \frac{1}{3}$ yd $= \frac{2}{3}$ yd	b. 6 ft = ________ yd 6 ft = 6 × 1 ft = 6 × ________ yd = ________ yd
c. 5 in = ________ ft	d. 14 in = ________ ft
e. 7 oz = ________ lb	f. 20 oz = ________ lb
g. 1 pt = ________ qt	h. 4 pt = ________ qt

2. Marty buys 12 ounces of granola.

   a. What fraction of a pound of granola did Marty buy?

   b. If a whole pound of granola costs $4, how much did Marty pay?

3. Sara and her dad visit Yo-Yo Yogurt again. This time, the scale says that Sara has 14 ounces of vanilla yogurt in her cup. Her father's yogurt weighs half as much. How many pounds of frozen yogurt did they buy altogether on this visit? Express your answer as a mixed number.

4. An art teacher uses 1 quart of blue paint each month. In one year, how many gallons of paint will she use?

Convert. Express the answer as a mixed number.

1. $2\frac{2}{3}$ ft = ____ in

1 foot = 12 inches

$$\begin{aligned} \mathbf{2\tfrac{2}{3}\ ft} &= \mathbf{2\tfrac{2}{3} \times 1\ ft} \\ &= \mathbf{2\tfrac{2}{3} \times 12\ in} \\ &= \mathbf{\tfrac{8}{3} \times 12\ in} \\ &= \mathbf{\tfrac{96}{3}\ in} \\ &= \mathbf{32\ in} \end{aligned}$$

I rename $2\frac{2}{3}$ as a fraction greater than 1, or an improper fraction, $\frac{8}{3}$. Then, I multiply.

2. $2\frac{7}{10}$ hr = _____ min

1 hour = 60 minutes

$$\begin{aligned} \mathbf{2\tfrac{7}{10}\ hr} &= \mathbf{2\tfrac{7}{10} \times 1\ hr} \\ &= \mathbf{2\tfrac{7}{10} \times 60\ min} \\ &= \mathbf{(2 \times 60\ min) + \left(\tfrac{7}{10} \times 60\ min\right)} \\ &= \mathbf{(120\ min) + (42\ min)} \\ &= \mathbf{162\ min} \end{aligned}$$

I can use the distributive property. I multiply 2 × 60 minutes and add that to the product of $\frac{7}{10}$ × 60 minutes.

3. Charlie buys $2\frac{1}{4}$ pounds of apples for a pie. He needs 50 ounces of apples for the pie. How many more pounds of apples does he need to buy?

I draw a whole tape diagram showing the total of 50 ounces of apples that Charlie needs for the pie.

50 oz

$2\frac{1}{4}$ **lb**　　**? lb**

I draw and label a part $2\frac{1}{4}$ pounds to show the apples Charlie bought.

I label the remaining part that Charlie needs with a question mark, to represent what I'm trying to find out.

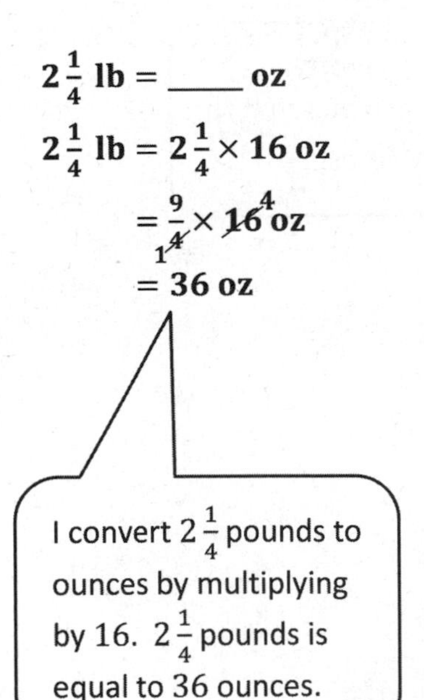

$2\frac{1}{4}\text{ lb} = \_\_\_\_\text{ oz}$

$2\frac{1}{4}\text{ lb} = 2\frac{1}{4} \times 16\text{ oz}$

$= \frac{9}{\cancel{4}_1} \times \cancel{16}^4\text{ oz}$

$= 36\text{ oz}$

I convert $2\frac{1}{4}$ pounds to ounces by multiplying by 16. $2\frac{1}{4}$ pounds is equal to 36 ounces.

$$\begin{array}{rrl} & \overset{4}{\cancel{5}}\ \overset{10}{\cancel{0}} & \text{oz} \\ - & 3\ \ 6 & \text{oz} \\ \hline & 1\ \ 4 & \text{oz} \end{array}$$

I subtract 36 ounces from the total of 50 ounces to find how many more ounces of apples Charlie needs to buy. The difference is 14 ounces.

$14\text{ oz} = \_\_\_\_\text{ lb}$

$14\text{ oz} = 14 \times 1\text{ oz}$

$= 14 \times \frac{1}{16}\text{ lb}$

$= \frac{14}{16}\text{ lb}$

$= \frac{7}{8}\text{ lb}$

Since the question asks how many more ***pounds*** does he need to buy, I convert 14 ounces to pounds.

Charlie needs to buy $\frac{7}{8}$ pound of apples.

Name ____________________ Date ____________

1. Convert. Show your work. Express your answer as a mixed number. The first one is done for you.

a. $2\frac{2}{3}$ yd = <u>8</u> ft $2\frac{2}{3}$ yd $= 2\frac{2}{3} \times 1$ yd $= 2\frac{2}{3} \times 3$ ft $= \frac{8}{3} \times 3$ ft $= \frac{24}{3}$ ft $= 8$ ft	b. $1\frac{1}{4}$ ft = ____________ yd $1\frac{1}{4}$ ft $= 1\frac{1}{4} \times 1$ ft $= 1\frac{1}{4} \times \frac{1}{3}$ yd $= \frac{5}{4} \times \frac{1}{3}$ yd $=$
c. $3\frac{5}{6}$ ft = ____________ in	d. $7\frac{1}{2}$ pt = ____________ qt
e. $4\frac{3}{10}$ hr = ____________ min	f. 33 months = ____________ years

2. Four members of a track team run a relay race in 165 seconds. How many minutes did it take them to run the race?

3. Horace buys $2\frac{3}{4}$ pounds of blueberries for a pie. He needs 48 ounces of blueberries for the pie. How many more pounds of blueberries does he need to buy?

4. Tiffany is sending a package that may not exceed 16 pounds. The package contains books that weigh a total of $9\frac{3}{8}$ pounds. The other items to be sent weigh $\frac{3}{5}$ the weight of the books. Will Tiffany be able to send the package?

Fill in the blanks

1. $\frac{3}{5} \times 1 = \frac{3}{5} \times \frac{\mathbf{6}}{\mathbf{6}} = \frac{18}{30}$

I think 3 times what is 18, and 5 times what is 30? The missing fraction must be $\frac{6}{6}$.

I know that any number times 1, or a fraction equal to 1, will be equal to the number itself.
$\frac{3}{5} = \frac{18}{30}$

2. Express each fraction as an equivalent decimal.

In order to write a fraction as a decimal, I can rename the denominator as a power of 10 (e.g., 10, 100, 1,000).
$\frac{1}{10} = 0.1$ $\frac{1}{100} = 0.01$ $\frac{1}{1{,}000} = 0.001$

a. $\frac{1}{4} \times \frac{\mathbf{25}}{\mathbf{25}} = \frac{\mathbf{25}}{\mathbf{100}} = \mathbf{0.25}$

I can rename $\frac{1}{4}$ as $\frac{25}{100}$, or 0.25.

I look at the denominator, 4, and it is a factor of 100 and 1,000.

b. $\frac{4}{5} \times \frac{\mathbf{2}}{\mathbf{2}} = \frac{\mathbf{8}}{\mathbf{10}} = \mathbf{0.8}$

I look at the denominator, 5, and it is a factor of 10, 100, and 1,000

c. $\frac{21}{20} \times \frac{\mathbf{5}}{\mathbf{5}} = \frac{\mathbf{105}}{\mathbf{100}} = \mathbf{1.05}$

Since $\frac{21}{20}$ is a fraction greater than 1, the equivalent decimal must also be greater than 1.

d. $3\frac{21}{50} \times \frac{\mathbf{2}}{\mathbf{2}} = \mathbf{3}\frac{\mathbf{42}}{\mathbf{100}} = \mathbf{3.42}$

Since $3\frac{21}{50}$ is a mixed number, the equivalent decimal must be greater than 1.

I look at the denominator, 50, and it is a factor of 100 and 1,000.

3. Vivian has $\frac{3}{4}$ of a dollar. She buys a lollipop for 59 cents. Change both numbers into decimals, and tell how much money Vivian has after paying for the lollipop.

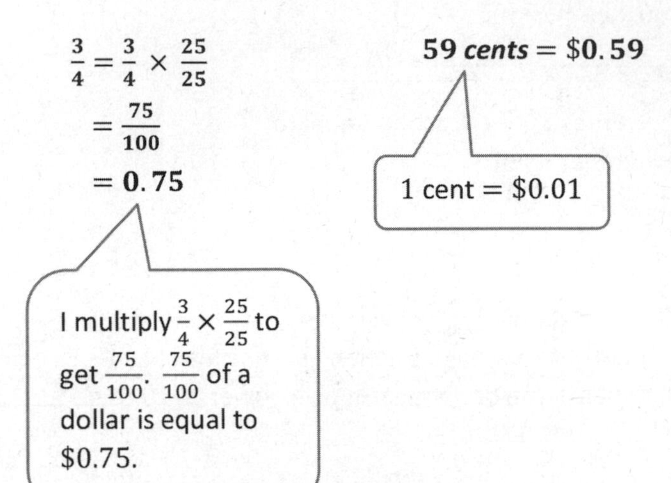

$$
\begin{array}{rr}
 & {}^{6}\ {}^{15} \\
 & \$0.\ \not7\ \not5 \\
- & \$0.\ 5\ 9 \\
\hline
 & \$0.\ 1\ 6
\end{array}
$$

I subtract $0.59 from $0.75 to find that Vivian has $0.16 left after paying for the lollipop.

Vivian has* $0.16 *left after paying for the lollipop.

Name ______________________________ Date ______________

1. Fill in the blanks.

a. $\frac{1}{3} \times 1 = \frac{1}{3} \times \frac{3}{3} = \frac{\quad}{9}$

b. $\frac{2}{3} \times 1 = \frac{2}{3} \times \frac{\quad}{\quad} = \frac{14}{21}$

c. $\frac{5}{2} \times 1 = \frac{5}{2} \times \frac{\quad}{\quad} = \frac{25}{\quad}$

d. Compare the first factor to the value of the product.

2. Express each fraction as an equivalent decimal. The first one is partially done for you.

a. $\frac{3}{4} \times \frac{25}{25} = \frac{3 \times 25}{4 \times 25} = \frac{\quad}{100} =$

b. $\frac{1}{4} \times \frac{25}{25} =$

c. $\frac{2}{5} \times \frac{\quad}{\quad} =$

d. $\frac{3}{5} \times \frac{\quad}{\quad} =$

e. $\frac{3}{20}$

f. $\frac{25}{20}$

g. $\frac{23}{25}$

h. $\frac{89}{50}$

i. $3\frac{11}{25}$

j. $5\frac{41}{50}$

3. $\frac{6}{8}$ is equivalent to $\frac{3}{4}$. How can you use this to help you write $\frac{6}{8}$ as a decimal? Show your thinking to solve.

4. A number multiplied by a fraction is not always smaller than the original number. Explain this and give at least two examples to support your thinking.

5. Elise has $\frac{3}{4}$ of a dollar. She buys a stamp that costs 44 cents. Change both numbers into decimals, and tell how much money Elise has after paying for the stamp.

1. Solve for the unknown. Rewrite each phrase as a multiplication sentence. Circle the scaling factor, and put a box around the factor naming the number of meters.

   a. $\frac{1}{2}$ as long as 8 meters = __4__ meters

   $\frac{1}{2} \times 8\text{ m} = 4\text{ m}$

   Half of 8 is 4, so 1 half of 8 *meters* is 4 *meters*.

   b. 8 times as long as $\frac{1}{2}$ meter = __4__ meters

   $8 \times \frac{1}{2}\text{ m} = 4\text{ m}$

   2 times 1 half is equal to 1. So 8 times 1 half (or 8 copies of 1 half) is equal to 4.

2. Draw a tape diagram to model each situation in Problem 1, and describe what happened to the number of meters when it was multiplied by the scaling factor.

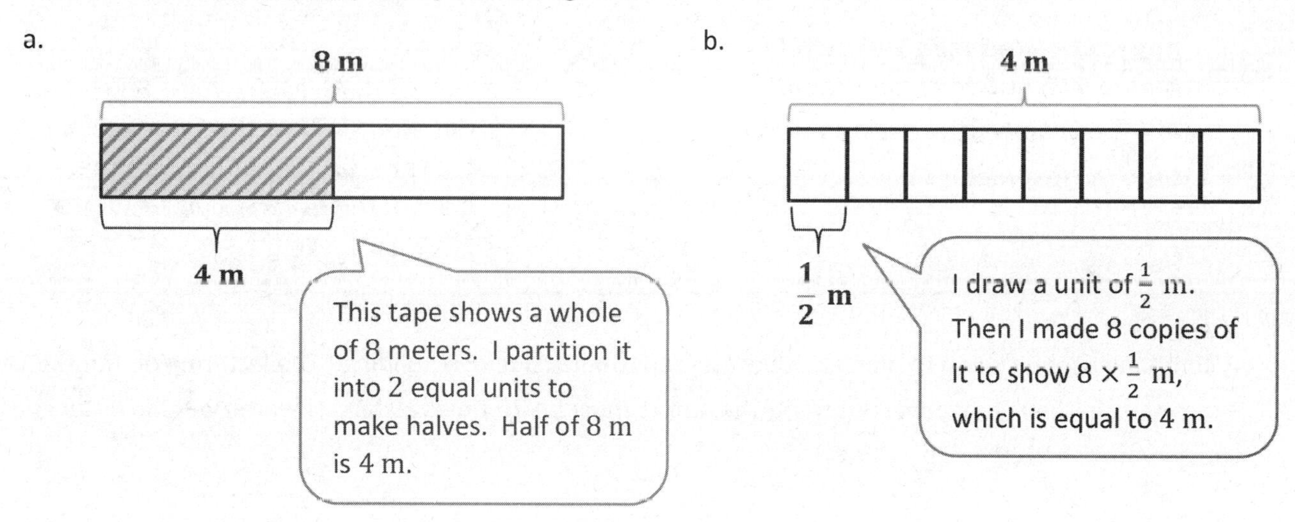

In part (a), the scaling factor $\frac{1}{2}$ ***is <u>less than 1</u>, so the number of meters <u>decreases</u>.***

In part (b), the scaling factor 8 is <u>greater than 1</u>, so the number of meters <u>increases</u>.

3. Look at the inequalities in each box. Choose a single fraction to write in all three blanks that would make all three number sentences true. Explain how you know.

a. $\frac{3}{4} \times \frac{4}{2} > \frac{3}{4}$ $2 \times \frac{4}{2} > 2$ $\frac{7}{5} \times \frac{4}{2} > \frac{7}{5}$

Any fraction greater than 1 will work. Multiplying by a factor greater than 1, like $\frac{4}{2}$, will make the product larger than the first factor shown.

Each of these inequalities shows that the expression on the left is greater than the value on the right. Therefore, I need to think of a scaling factor that is greater than 1, like $\frac{4}{2}$.

b. $\frac{3}{4} \times \frac{1}{3} < \frac{3}{4}$ $2 \times \frac{1}{3} < 2$ $\frac{7}{5} \times \frac{1}{3} < \frac{7}{5}$

Any fraction less than 1 will work. Multiplying by a factor less than 1, like $\frac{1}{3}$, will make the product smaller than the first factor shown.

Each of these inequalities shows that the expression on the left is less than the value on the right. Therefore, I need to think of a scaling factor that is less than 1, like $\frac{1}{3}$.

4. A company uses a sketch to plan an advertisement on the side of a building. The lettering on the sketch is $\frac{3}{4}$ inch tall. In the actual advertisement, the letters must be 20 times as tall. How tall will the letters be on the actual advertisement?

$$20 \times \frac{3}{4}$$
$$= \frac{20 \times 3}{4}$$
$$= \frac{60}{4}$$
$$= 15$$

The letters will be 15 inches tall.

The letters on the sketch have been scaled down to fit on the page; therefore, the letters on the actual advertisement will be larger. In order to find out how large the actual letters will be, I multiply 20 by $\frac{3}{4}$ inch.

Name ____________________ Date ____________

1. Solve for the unknown. Rewrite each phrase as a multiplication sentence. Circle the scaling factor and put a box around the number of meters.

   a. $\frac{1}{3}$ as long as 6 meters = ______ meter(s)

   b. 6 times as long as $\frac{1}{3}$ meter = ______ meter(s)

2. Draw a tape diagram to model each situation in Problem 1, and describe what happened to the number of meters when it was multiplied by the scaling factor.

   a.

   b.

3. Fill in the blank with a numerator or denominator to make the number sentence true.

   a. $5 \times \frac{\_\_}{3} > 5$

   b. $\frac{6}{\_\_} \times 12 < 12$

   c. $4 \times \frac{\_\_}{5} = 4$

4. Look at the inequalities in each box. Choose a single fraction to write in all three blanks that would make all three number sentences true. Explain how you know.

   a. $\frac{2}{3} \times \_\_\_\_ > \frac{2}{3}$ $\quad 4 \times \_\_\_\_ > 4$ $\quad \frac{5}{3} \times \_\_\_\_ > \frac{5}{3}$

   b. $\frac{2}{3} \times \_\_\_\_ < \frac{2}{3}$ $\quad 4 \times \_\_\_\_ < 4$ $\quad \frac{5}{3} \times \_\_\_\_ < \frac{5}{3}$

5. Write a number in the blank that will make the number sentence true.

   a. 3 × ______ < 1

   b. Explain how multiplying by a whole number can result in a product less than 1.

6. In a sketch, a fountain is drawn $\frac{1}{4}$ yard tall. The actual fountain will be 68 times as tall. How tall will the fountain be?

7. In blueprints, an architect's firm drew everything $\frac{1}{24}$ of the actual size. The windows will actually measure 4 ft by 6 ft and doors measure 12 ft by 8 ft. What are the dimensions of the windows and the doors in the drawing?

1. Sort the following expressions by rewriting them in the table.

[13.89] × 1.004	[602] × 0.489	[102.03] × 4.015
[0.3] × 0.069	[0.72] × 1.24	[0.2] × 0.1

The product is less than the boxed number:	The product is greater than the boxed number:
[0.3] × 0.069 [602] × 0.489 [0.2] × 0.1	[13.89] × 1.004 [0.72] × 1.24 [102.03] × 4.015

All of the expressions in this column have a boxed number that is multiplied by a **scaling factor less than 1** (e.g., 0.069 and 0.1). Therefore, the product will be less than the boxed number.

All of the expressions in this column have a boxed number that is multiplied by a **scaling factor more than 1** (e.g., 1.004 and 4.015). Therefore, the product will be greater than the boxed number.

2. Write a statement using one of the following phrases to compare the value of the expressions.

is slightly more than *is a lot more than* *is slightly less than* *is a lot less than*

a. 4×0.988 ***is slightly less than*** 4

In this example, the product of 4×0.988 is being compared to the factor 4. Since the scaling factor, 0.988, is less than 1, the product will be less than 4. However, since the scaling factor, 0.988, is just ***slightly*** less than 1, the factor will also be ***slightly*** less than 4.

b. 1.05×0.8 ***is slightly more than*** 0.8

c. $1{,}725 \times 0.013$ ***is a lot less than*** $1{,}725$

d. 89.001×1.3 ***is a lot more than*** 1.3

In this example, the product of 89.001×1.3 is being compared to the factor 1.3. Since the scaling factor, 89.001, is more than 1, the product will be more than 1.3. However, since the scaling factor, 89.001, is ***a lot more*** than 1, the product will also be ***a lot more*** than 1.3.

3. During science class, Teo, Carson, and Dhakir measure the length of their bean sprouts. Carson's sprout is 0.9 times the length of Teo's, and Dhakir's is 1.08 times the length of Teo's. Whose bean sprout is the longest? The shortest?

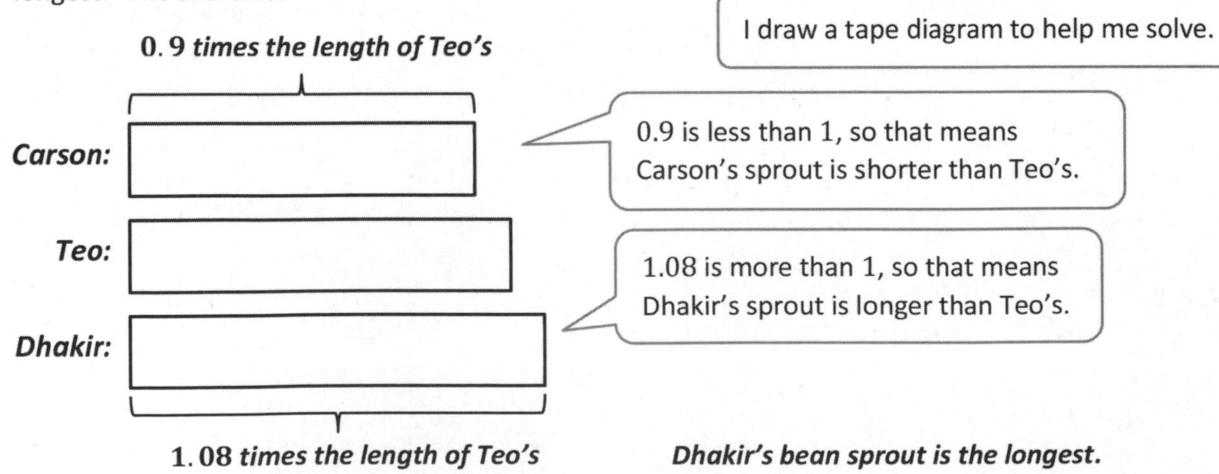

Dhakir's bean sprout is the longest.

Carson's bean sprout is the shortest.

Name ______________________________ Date ________________

1.

a. Sort the following expressions by rewriting them in the table.

The product is less than the boxed number:	The product is greater than the boxed number:

[12.5] × 1.989 [828] × 0.921 [321.46] × 1.26

[0.007] × 1.02 [2.16] × 1.11 [0.05] × 0.1

b. What do the expressions in each column have in common?

2. Write a statement using one of the following phrases to compare the value of the expressions. Then, explain how you know.

is slightly more than ***is a lot more than*** ***is slightly less than*** ***is a lot less than***

a. 14 × 0.999 ______________________ 14

b. 1.01 × 2.06 ______________________ 2.06

c. 1,955 × 0.019 ______________________ 1,955

d. Two thousand × 1.0001 ______________________ two thousand

e. Two-thousandths × 0.911 ______________________ two-thousandths

3. Rachel is 1.5 times as heavy as her cousin, Kayla. Another cousin, Jonathan, weighs 1.25 times as much as Kayla. List the cousins, from lightest to heaviest, and explain your thinking.

4. Circle your choice.

a. $a \times b > a$

For this statement to be true, b must be **greater than 1** **less than 1**

Write two expressions that support your answer. Be sure to include one decimal example.

b. $a \times b < a$

For this statement to be true, b must be **greater than 1** **less than 1**

Write two expressions that support your answer. Be sure to include one decimal example.

1. A tube contains 28 mL of medicine. If each dose is $\frac{1}{8}$ of the tube, how many milliliters is each dose? Express your answer as a decimal.

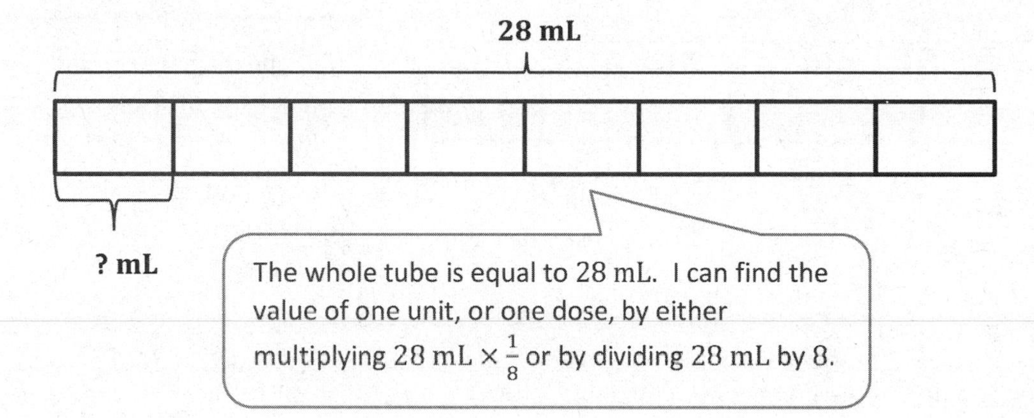

8 units **= 28 mL**

1 unit **= 28 mL ÷ 8**

$= \frac{28}{8}$ mL

$= 3\frac{4}{8}$ mL

$= 3\frac{1}{2}$ mL

Now I know that each dose is $3\frac{1}{2}$ mL, but the problem asks me to express my answer as a decimal. I'll need to find a fraction that is equal to $\frac{1}{2}$ and has a denominator of 10, 100, or 1,000.

I can multiply the fraction $\frac{1}{2}$ by $\frac{5}{5}$ to create an equivalent fraction with 10 as the denominator. Then I'll be able to express $3\frac{1}{2}$ as a decimal.

Each dose is $3\frac{1}{2}$ **mL.**

$3\frac{1}{2} \times \frac{5}{5} = 3\frac{5}{10} = 3.5$

Each dose is **3.5 mL.**

Note: Some students may recognize that the fraction $\frac{1}{2}$ is equal to 0.5 without showing any work. Encourage your child to show the amount of work that is necessary to be successful. If your child can do basic calculations mentally, allow him or her to do so!

2. A clothing factory uses 1,275.2 meters of cloth a week to make shirts. How much cloth is needed to make $3\frac{3}{5}$ times as many shirts?

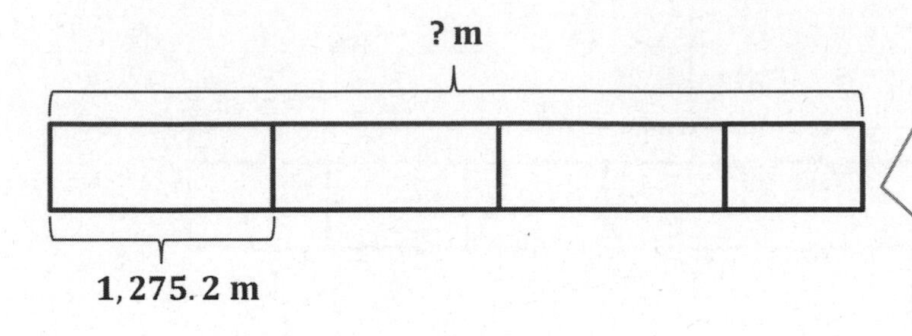

My tape diagram reminds me that I can use the distributive property to solve. I can multiply $1{,}275\frac{2}{10}$ by 3 first, to find out what 3 times as many shirts is. Then I can multiply by $\frac{3}{5}$ to find out what $\frac{3}{5}$ as many shirts is.

$$1{,}275.2 \text{ m} = 1{,}275\frac{2}{10} \text{ m}$$

I can rename 2 tenths meter as a fraction.

$$
\begin{aligned}
1{,}275\frac{2}{10} \times 3\frac{3}{5} &= \left(1{,}275\frac{2}{10} \times 3\right) + \left(1{,}275\frac{2}{10} \times \frac{3}{5}\right) \\
&= \left(3{,}825\frac{6}{10}\right) + \left(\frac{12{,}752}{10} \times \frac{3}{5}\right) \\
&= \left(3{,}825\frac{6}{10}\right) + \left(\frac{12{,}752 \times 3}{10 \times 5}\right) \\
&= \left(3{,}825\frac{6}{10}\right) + \left(\frac{38{,}256}{50}\right) \\
&= \left(3{,}825\frac{6}{10}\right) + \left(765\frac{6}{50}\right) \\
&= \left(3{,}825\frac{60}{100}\right) + \left(765\frac{12}{100}\right) \\
&= 4{,}590\frac{72}{100} \\
&= 4{,}590.72
\end{aligned}
$$

In order to add, I make like units, or find common denominators. I'll rename each fraction using hundredths, so I can easily express my final answer as a decimal.

I can rename $\frac{72}{100}$ as 0.72 to express my final answer as a decimal.

4,590.72 ***meters of cloth are needed to make the shirts.***

3. There are $\frac{3}{4}$ as many boys as girls in a class of fifth graders. If there are 35 students in the class, how many are girls?

I draw a tape to represent the number of girls in the class.

I partition it into 4 equal units to make fourths.

Girls:

Boys:

35

I can think about what my tape diagram is showing. There are a total of 7 units, and those 7 units are equal to a total of 35 students. In order to find out how many girls there are, I need to know the value of 1 unit.

Since there are $\frac{3}{4}$ as many boys as girls, I draw a tape to represent the number of boys that is $\frac{3}{4}$ as long as the tape for the number of girls.

$7 \text{ units} = 35$

$1 \text{ unit} = 35 \div 7$

$1 \text{ unit} = 5$

$4 \text{ units} = 4 \times 5 = 20$

There are 20 girls in the class.

If each unit is equal to 5 students and there are 4 units representing the girls, I can multiply to find the number of girls in the class.

Name ________________________________ Date ____________________

1. Jesse takes his dog and cat for their annual vet visit. Jesse's dog weighs 23 pounds. The vet tells him his cat's weight is $\frac{5}{8}$ as much as his dog's weight. How much does his cat weigh?

2. An image of a snowflake is 1.8 centimeters wide. If the actual snowflake is $\frac{1}{8}$ the size of the image, what is the width of the actual snowflake? Express your answer as a decimal.

3. A community bike ride offers a short 5.7-mile ride for children and families. The short ride is followed by a long ride, $5\frac{2}{3}$ times as long as the short ride, for adults. If a woman bikes the short ride with her children and then the long ride with her friends, how many miles does she ride altogether?

4. Sal bought a house for $78,524.60. Twelve years later he sold the house for $2\frac{3}{4}$ times as much. What was the sale price of the house?

5. In the fifth grade at Lenape Elementary School, there are $\frac{4}{5}$ as many students who do not wear glasses as those who do wear glasses. If there are 60 students who wear glasses, how many students are in the fifth grade?

6. At a factory, a mechanic earns $17.25 an hour. The president of the company earns $6\frac{2}{3}$ times as much for each hour he works. The janitor at the same company earns $\frac{3}{5}$ as much as the mechanic. How much does the company pay for all three employees' wages for one hour of work?

1. Draw a tape diagram and a number line to solve.

$2 \div \frac{1}{2}$

I can think about this division expression by asking "How many halves are in 2 wholes?"

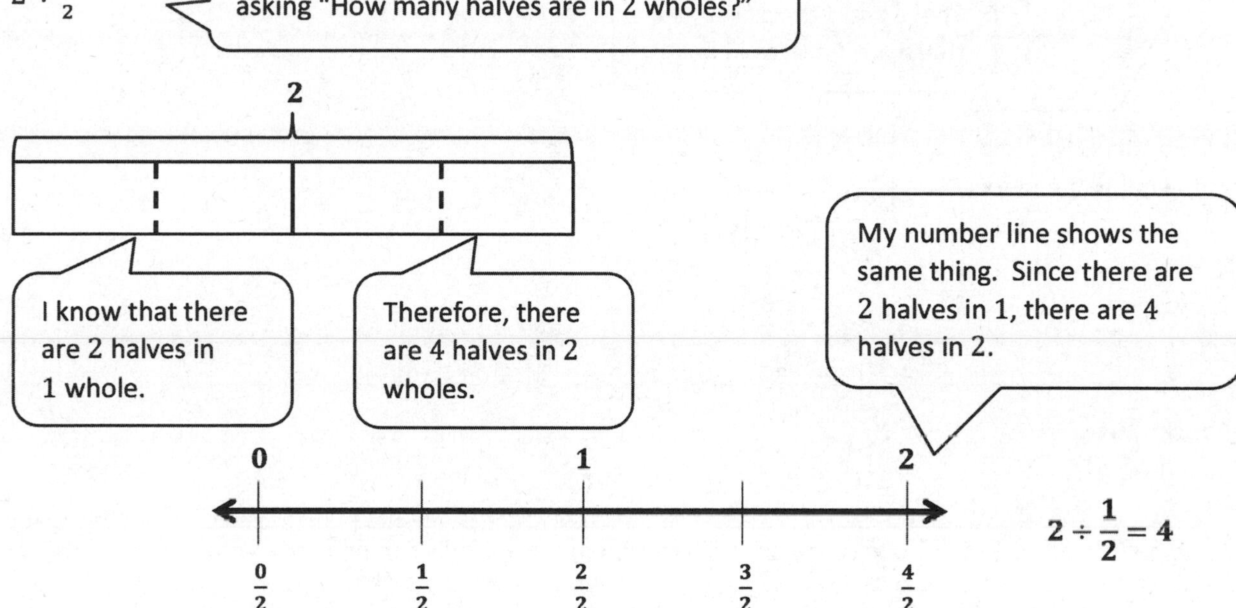

$2 \div \frac{1}{2} = 4$

$2 \div \frac{1}{2}$

I can also think about this division expression by asking "2 is ***half*** of what?" or "If 2 is ***half***, what is the whole?"

I draw a unit of 2.

And since 2 is half, I draw another unit of 2.

My number line shows the same thing. If 2 is half, 4 is the whole.

2

4

0 1 2 3 4

$\frac{1}{2}$ ***of the total***

$\frac{1}{2}$ ***of the total***

Therefore, if 2 is half, 4 is the whole!

2. Divide. Then multiply to check.

$2 \div \frac{1}{3}$

I can think, "How many thirds are in 2?"
There are 3 thirds in 1, so there are 6 thirds in 2.

Or I can think, "If 2 is a third, what is the whole?"

$2 \div \frac{1}{3} = 6$

Check: $6 \times \frac{1}{3} = \frac{6 \times 1}{3} = \frac{6}{3} = 2$

3. A recipe for rolls calls for $\frac{1}{4}$ cup of sugar. How many batches of rolls can be made with 2 cups of sugar?

This problem is asking me to find how many fourths are in 2.

There are a total of 2 cups of sugar.

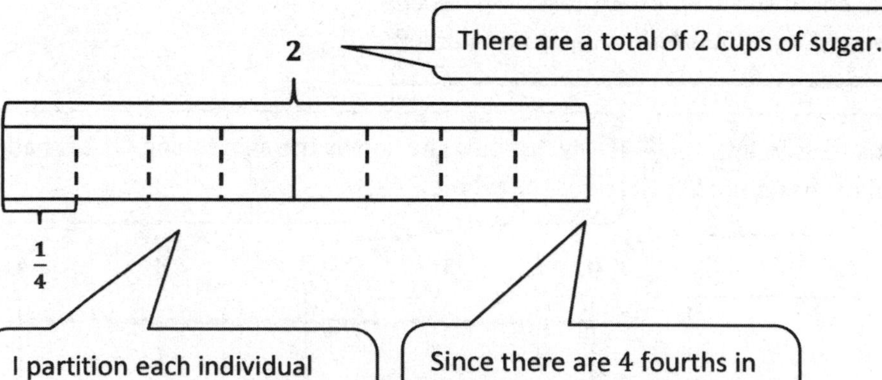

I partition each individual cup of sugar into 4 equal units, called fourths.

Since there are 4 fourths in 1 cup, there are 8 fourths in 2 cups.

$2 \div \frac{1}{4} = 8$

8 batches of rolls can be made with 2 cups of sugar.

Name ______________________________ Date ______________

1. Draw a tape diagram and a number line to solve. Fill in the blanks that follow.

a. $3 \div \frac{1}{3} =$ ________

There are ____ thirds in 1 whole.

There are ____ thirds in 3 wholes.

If 3 is $\frac{1}{3}$, what is the whole? ______

b. $3 \div \frac{1}{4} =$ ________

There are ____ fourths in 1 whole.

There are ____ fourths in ___ wholes.

If 3 is $\frac{1}{4}$, what is the whole? ______

c. $4 \div \frac{1}{3} =$ ________

There are ____ thirds in 1 whole.

There are ____ thirds in __ wholes.

If 4 is $\frac{1}{3}$, what is the whole? ______

d. $5 \div \frac{1}{4} =$ ________

There are ____ fourths in 1 whole.

There are ____ fourths in __ wholes.

If 5 is $\frac{1}{4}$, what is the whole? ______

2. Divide. Then, multiply to check.

a. $2 \div \frac{1}{4}$	b. $6 \div \frac{1}{2}$	c. $5 \div \frac{1}{4}$	d. $5 \div \frac{1}{8}$
e. $6 \div \frac{1}{3}$	f. $3 \div \frac{1}{6}$	g. $6 \div \frac{1}{5}$	h. $6 \div \frac{1}{10}$

3. A principal orders 8 sub sandwiches for a teachers' meeting. She cuts the subs into thirds and puts the mini-subs onto a tray. How many mini-subs are on the tray?

4. Some students prepare 3 different snacks. They make $\frac{1}{8}$ pound bags of nut mix, $\frac{1}{4}$ pound bags of cherries, and $\frac{1}{6}$ pound bags of dried fruit. If they buy 3 pounds of nut mix, 5 pounds of cherries, and 4 pounds of dried fruit, how many of each type of snack bag will they be able to make?

1. Solve and support your answer with a model or tape diagram. Write your quotient in the blank.

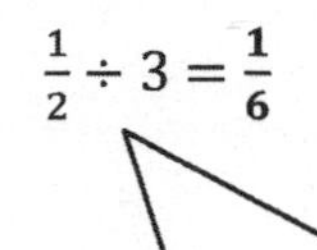

I can think of this expression as "One half of a pan of brownies is shared equally with 3 people. How much of the pan does each person get?"

1

I can draw a pan of brownies and shade the $\frac{1}{2}$ of a pan that will be shared.

1

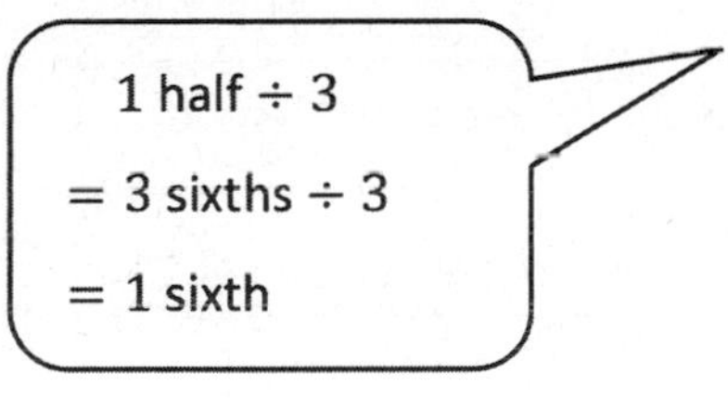

In order to share the brownies with 3 people equally, I partition it into 3 equal parts. I do the same for the other half of the pan so that I can see equal units. Each person will get $\frac{1}{6}$ of the pan of brownies.

2. Divide. Then, multiply to check.

$\frac{1}{4} \div 5$

$\frac{5}{20} \div 5 =$ **5 *twentieths*** $\div$ **5 = 1 *twentieth*** $= \frac{1}{20}$

I can visualize a tape diagram. In my mind, I can see 1 fourth being partitioned into 5 equal units. Now, instead of seeing fourths, the model is showing twentieths.

I know that 5 ÷ 5 is equal to 1.
Therefore, 5 *twentieths* ÷ 5 = 1 *twentieth*, or $\frac{1}{20}$.

Check: $\frac{1}{20} \times 5 = \frac{5}{20} = \frac{1}{4}$

I check my answer by multiplying the quotient, $\frac{1}{20}$, and the divisor, 5, to get $\frac{1}{4}$.

Since Tim read $\frac{4}{5}$ of the book, it means he has $\frac{1}{5}$ left to read.
$1 - \frac{4}{5} = \frac{1}{5}$

3. Tim has read $\frac{4}{5}$ of his book. He finishes the book by reading the same amount each night for 3 nights.

   a. What fraction of the book does he read on each of the 3 nights?

   $$\frac{1}{5} \div 3 = \frac{3}{15} \div 3 = \frac{1}{15}$$

   I can rename $\frac{1}{5}$ as $\frac{3}{15}$. Then, I divide.
   3 fifteenths $\div 3 =$ 1 fifteenth, or $\frac{1}{15}$.

   He reads $\frac{1}{15}$ of the book on each of the 3 nights.

   b. If he reads 6 pages on each of the 3 nights, how long is the book?

   1 *unit* = 6 *pages*

   15 *units* = 15 × 6 *pages* = 90 *pages*

   Tim reads $\frac{1}{15}$, or 6 pages, each night.
   So $\frac{1}{15}$ or 1 unit is equal to 6 pages.

   The book has 90 pages.

   The whole book is equal to $\frac{15}{15}$, or 15 units.
   So I multiply 15 times 6.

Name ____________________________ Date ____________

1. Solve and support your answer with a model or tape diagram. Write your quotient in the blank.

a. $\frac{1}{2} \div 4 =$ ______

b. $\frac{1}{3} \div 6 =$ ______

c. $\frac{1}{4} \div 3 =$ ______

d. $\frac{1}{5} \div 2 =$ ______

2. Divide. Then, multiply to check.

a. $\frac{1}{2} \div 10$	b. $\frac{1}{4} \div 10$	c. $\frac{1}{3} \div 5$	d. $\frac{1}{5} \div 3$
e. $\frac{1}{8} \div 4$	f. $\frac{1}{7} \div 3$	g. $\frac{1}{10} \div 5$	h. $\frac{1}{5} \div 20$

3. Teams of four are competing in a quarter-mile relay race. Each runner must run the same exact distance. What is the distance each teammate runs?

4. Solomon has read $\frac{1}{3}$ of his book. He finishes the book by reading the same amount each night for 5 nights.

   a. What fraction of the book does he read each of the 5 nights?

   b. If he reads 14 pages on each of the 5 nights, how long is the book?

1. Owen ordered 2 mini cakes for a birthday party. The cakes were sliced into fifths. How many slices were there? Draw a picture to support your response.

I draw a tape diagram and label 2 for the 2 mini cakes.

I can think, "How many fifths are in 2?"

2

I cut each cake into 5 equal units and get a total of 10 units.

5 fifths in 1 cake

10 fifths in 2 cakes

$2 \div \frac{1}{5} = 10$

There were 10 slices.

2. Alex has $\frac{1}{8}$ of a pizza left over. He wants to give the leftover pizza to 3 friends to share equally. What fraction of the original pizza will each friend receive? Draw a picture to support your response.

I draw a tape diagram and label it 1 to represent the whole pizza. I cut it into 8 equal units and shade 1 unit to represent the $\frac{1}{8}$ that Alex has.

Three friends are sharing $\frac{1}{8}$ of a pizza. I'll divide $\frac{1}{8}$ by 3 to find how much each friend will receive.

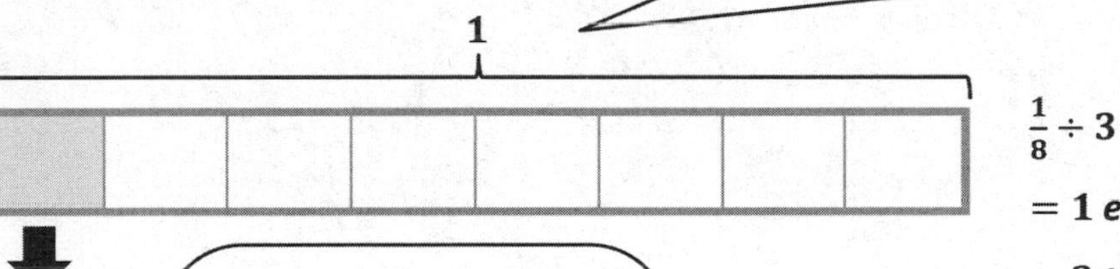

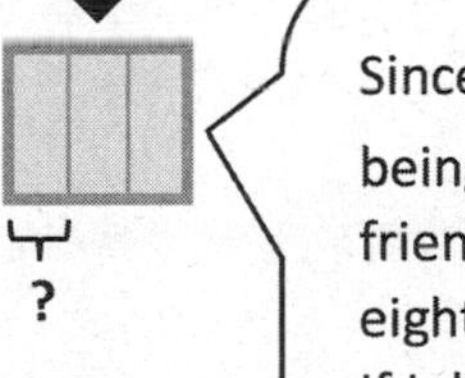

Since the $\frac{1}{8}$ of a pizza is being shared by 3 friends, I partition the eighth into 3 equal parts. If I did that with the other $\frac{7}{8}$, that would make a total of 24 units.

$\frac{1}{8} \div 3$

$= 1$ ***eighth*** $\div 3$

$= 3$ ***twenty-fourths*** $\div 3$

$= 1$ ***twenty-fourth***

Each friend will receive $\frac{1}{24}$ ***of a pizza.***

One eighth is equal to 3 twenty-fourths. Three twenty-fourths divided by 3 is equal to 1 twenty-fourth.

Name ______________________________ Date ________________

1. Kelvin ordered four pizzas for a birthday party. The pizzas were cut in eighths. How many slices were there? Draw a picture to support your response.

2. Virgil has $\frac{1}{6}$ of a birthday cake left over. He wants to share the leftover cake with 3 friends. What fraction of the original cake will each of the 4 people receive? Draw a picture to support your response.

3. A pitcher of water contains $\frac{1}{4}$ liters of water. The water is poured equally into 5 glasses.

   a. How many liters of water are in each glass? Draw a picture to support your response.

b. Write the amount of water in each glass in milliliters.

4. Drew has 4 pieces of rope 1 meter long each. He cuts each rope into fifths.

   a. How many fifths will he have after cutting all the ropes?

   b. How long will each of the fifths be in centimeters?

5. A container is filled with blueberries. $\frac{1}{6}$ of the blueberries is poured equally into two bowls.

   a. What fraction of the blueberries is in each bowl?

   b. If each bowl has 6 ounces of blueberries in it, how many ounces of blueberries were in the full container?

   c. If $\frac{1}{5}$ of the remaining blueberries is used to make muffins, how many pounds of blueberries are left in the container?

My story problem has to be about 4 meters of string.

1. Create and solve a division story problem about 4 meters of string that is modeled by the tape diagram below.

The whole or dividend is 4 meters, and it is being cut into units of $\frac{1}{3}$ meter. One third is the divisor.

4

$\frac{1}{3}$	$\frac{1}{3}$	. . .	$\frac{1}{3}$

? thirds

How many thirds are in 4? I can solve by dividing, $4 \div \frac{1}{3}$.

Allison has 4 meters of string. She cuts each meter equally into thirds. How many thirds will she have altogether?

$4 \div \frac{1}{3} = 12$

Allison will have 12 thirds.

Since there are 3 thirds in 1, $2 = 6$ thirds, $3 = 9$ thirds, and $4 = 12$ thirds. Therefore, 4 divided by $\frac{1}{3}$ is equal to 12.

2. Create and solve a story problem about $\frac{1}{3}$ pound of peanuts that is modeled by the tape diagram below.

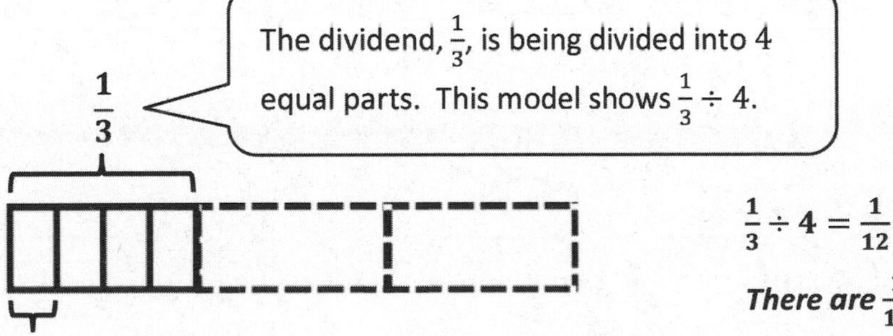

$\frac{1}{3} \div 4 = \frac{1}{12}$

There are $\frac{1}{12}$ pound of peanuts in each bag.

Juanita bought $\frac{1}{3}$ pound of peanuts. She splits the peanuts equally into 4 bags. How many pounds of peanuts are in each bag?

3. Draw a tape diagram and create a word problem for the following expressions, and then solve.

$2 \div \frac{1}{5} = \mathbf{10}$

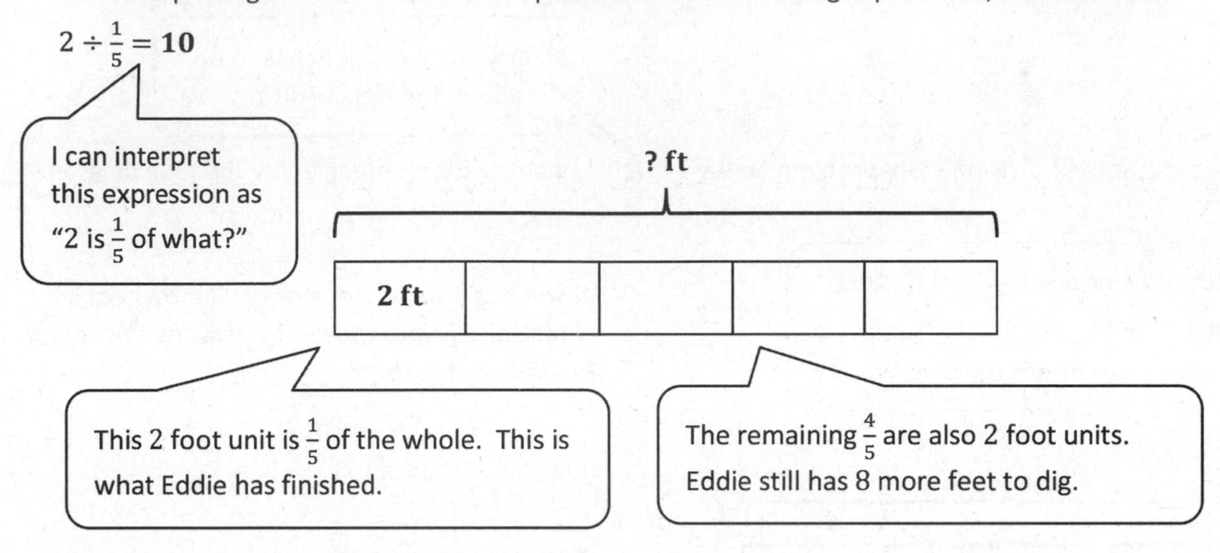

After digging a tunnel 2 feet long, Eddie had finished $\frac{1}{5}$ of the tunnel. How long will the tunnel be when Eddie is done?

The tunnel will be 10 feet long.

Name ______________________________ Date ______________

1. Create and solve a division story problem about 7 feet of rope that is modeled by the tape diagram below.

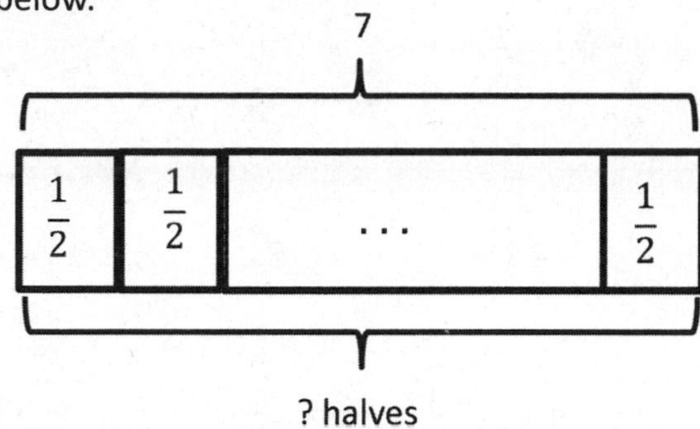

2. Create and solve a story problem about $\frac{1}{3}$ pound of flour that is modeled by the tape diagram below.

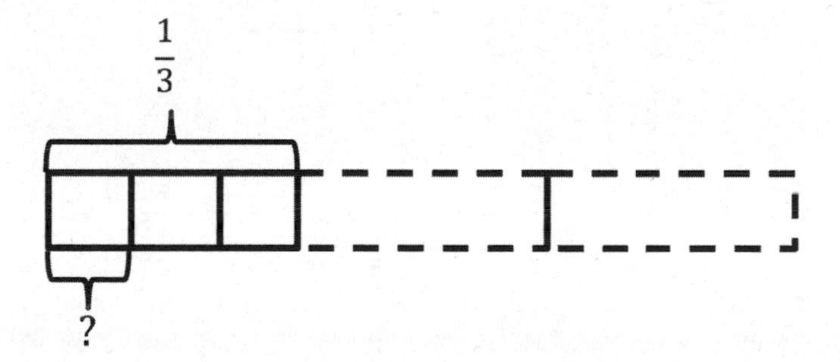

3. Draw a tape diagram and create a word problem for the following expressions. Then, solve and check.

   a. $2 \div \frac{1}{4}$

   b. $\frac{1}{4} \div 2$

   c. $\frac{1}{3} \div 5$

   d. $3 \div \frac{1}{10}$

1. Divide. Rewrite each expression as a division sentence with a fraction divisor, and fill in the blanks.

   a. $4 \div 0.1 = \mathbf{4 \div \frac{1}{10} = 40}$

   There are **10** tenths in 1 whole.

   There are **40** tenths in 4 wholes.

   b. $3.5 \div 0.1 = \mathbf{3.5 \div \frac{1}{10} = 35}$

   There are 10 tenths in 1, so there are 30 tenths in 3.

   There are **30** tenths in 3 wholes.

   There are **5** tenths in 5 tenths.

   There are **35** tenths in 3.5.

   c. $5 \div 0.01 = \mathbf{5 \div \frac{1}{100} = 500}$

   There are **100** hundredths in 1 whole.

   There are **500** hundredths in 5 whole.

   d. $2.7 \div 0.01 = \mathbf{2.7 \div \frac{1}{100} = 270}$

   There are 100 hundredths in 1, so there are 200 hundredths in 2.

   There are **200** hundredths in 2 wholes.

   There are **70** hundredths in 7 tenths.

   There are 10 hundredths in 1 tenth, so there are 70 hundredths in 7 tenths.

   There are **270** hundredths in 2.7.

2. Divide.

a. $35 \div 0.1$

$= \mathbf{35 \div \frac{1}{10}}$

$= \mathbf{350}$

I know that there are 10 tenths in 1 and 100 tenths in 10. So there are 350 tenths in 35.

b. $1.9 \div 0.1$

$= \mathbf{1.9 \div \frac{1}{10}}$

$= \mathbf{19}$

I can decompose 1.9 into 1 one 9 tenths. There are 10 tenths in 1, and 9 tenths in 9 tenths. Therefore, there are 19 tenths in 1.9.

c. $3.76 \div 0.01$

$= \mathbf{3.76 \div \frac{1}{100}}$

$= \mathbf{376}$

I can decompose 3.76 into 3 ones 7 tenths 6 hundredths. 3 ones = 300 hundredths, 7 tenths = 70 hundredths, and 6 hundredths = 6 hundredths.

Name ______________________________ Date ____________________

1. Divide. Rewrite each expression as a division sentence with a fraction divisor, and fill in the blanks. The first one is done for you.

Example: $4 \div 0.1 = 4 \div \frac{1}{10} = 40$

There are **10** tenths in 1 whole.

There are **40** tenths in 4 wholes.

a. $9 \div 0.1$

There are ______ tenths in 1 whole.

There are ______ tenths in 9 wholes.

b. $6 \div 0.1$

There are ______ tenths in 1 whole.

There are ______ tenths in 6 wholes.

c. $3.6 \div 0.1$

There are ______ tenths in 3 wholes.

There are ______ tenths in 6 tenths.

There are ______ tenths in 3.6.

d. $12.8 \div 0.1$

There are ______ tenths in 12 wholes.

There are ______ tenths in 8 tenths.

There are ______ tenths in 12.8.

e. $3 \div 0.01$

There are ______ hundredths in 1 whole.

There are ______ hundredths in 3 wholes.

f. $7 \div 0.01$

There are ______ hundredths in 1 whole.

There are ______ hundredths in 7 wholes.

g. $4.7 \div 0.01$

There are ______ hundredths in 4 wholes.

There are ______ hundredths in 7 tenths.

There are ______ hundredths in 4.7.

h. $11.3 \div 0.01$

There are ______ hundredths in 11 wholes.

There are ______ hundredths in 3 tenths.

There are ______ hundredths in 11.3.

2. Divide.

a. $2 \div 0.1$	b. $23 \div 0.1$	c. $5 \div 0.01$
d. $7.2 \div 0.1$	e. $51 \div 0.01$	f. $31 \div 0.1$
g. $231 \div 0.1$	h. $4.37 \div 0.01$	i. $24.5 \div 0.01$

3. Giovanna is charged $0.01 for each text message she sends. Last month, her cell phone bill included a $12.60 charge for text messages. How many text messages did Giovanna send?

4. Geraldine solved a problem: $68.5 \div 0.01 = 6{,}850$.

   Ralph said, "This is wrong because a quotient can't be greater than the whole you start with. For example, $8 \div 2 = 4$ and $250 \div 5 = 50$." Who is correct? Explain your thinking.

5. The price for an ounce of gold on September 23, 2013, was $1,326.40. A group of 10 friends decide to equally share the cost of 1 ounce of gold. How much money will each friend pay?

1. Rewrite the division expression as a fraction and divide.

a. $6.3 \div 0.9 = \frac{\mathbf{6.3}}{\mathbf{0.9}}$

$= \frac{\mathbf{6.3 \times 10}}{\mathbf{0.9 \times 10}}$

$= \frac{\mathbf{63}}{\mathbf{9}}$

$= \mathbf{7}$

I can multiply this fraction by 1, or $\frac{10}{10}$, to get a denominator that is a whole number.

After multiplying by $\frac{10}{10}$, the division expression is 63 divided by 9.

b. $6.3 \div 0.09 = \frac{\mathbf{6.3}}{\mathbf{0.09}}$

$= \frac{\mathbf{6.3 \times 100}}{\mathbf{0.09 \times 100}}$

$= \frac{\mathbf{630}}{\mathbf{9}}$

$= \mathbf{70}$

I can multiply this fraction by 1, or $\frac{100}{100}$, to get a denominator that is a whole number.

c. $4.8 \div 1.2 = \frac{\mathbf{4.8}}{\mathbf{1.2}}$

$= \frac{\mathbf{4.8 \times 10}}{\mathbf{1.2 \times 10}}$

$= \frac{\mathbf{48}}{\mathbf{12}}$

$= \mathbf{4}$

d. $0.48 \div 0.12 = \frac{\mathbf{0.48}}{\mathbf{0.12}}$

$= \frac{\mathbf{0.48 \times 100}}{\mathbf{0.12 \times 100}}$

$= \frac{\mathbf{48}}{\mathbf{12}}$

$= \mathbf{4}$

2. Mr. Huynh buys 2.4 kg of flour for his bakery.

a. If he pours 0.8 kg of flour into separate bags, how many bags of flour can he make?

I can divide 2.4 kg by 0.8 kg to find the number of bags of flour he can make.

$$\mathbf{2.4 \div 0.8 = \frac{2.4}{0.8} = \frac{2.4 \times 10}{0.8 \times 10} = \frac{24}{8} = 3}$$

24 divided by 8 is equal to 3.

He can make 3 bags of flour.

b. If he pours 0.4 kg of flour into separate bags, how many bags of flour can he make?

$$\mathbf{2.4 \div 0.4 = \frac{2.4}{0.4} = \frac{2.4 \times 10}{0.4 \times 10} = \frac{24}{4} = 6}$$

He can make 6 bags of flour.

Name ______________________________ Date ________________

1. Rewrite the division expression as a fraction and divide. The first two have been started for you.

a. $2.4 \div 0.8 = \frac{2.4}{0.8}$ $= \frac{2.4 \times 10}{0.8 \times 10}$ $= \frac{24}{8}$ $=$	b. $2.4 \div 0.08 = \frac{2.4}{0.08}$ $= \frac{2.4 \times 100}{0.08 \times 100}$ $= \frac{240}{8}$ $=$
c. $4.8 \div 0.6$	d. $0.48 \div 0.06$
e. $8.4 \div 0.7$	f. $0.84 \div 0.07$

g. $4.5 \div 1.5$	h. $0.45 \div 0.15$
i. $14.4 \div 1.2$	j. $1.44 \div 0.12$

2. Leann says $18 \div 6 = 3$, so $1.8 \div 0.6 = 0.3$ and $0.18 \div 0.06 = 0.03$. Is Leann correct? Explain how to solve these division problems.

3. Denise is making bean bags. She has 6.4 pounds of beans.

   a. If she makes each bean bag 0.8 pounds, how many bean bags will she be able to make?

   b. If she decides instead to make mini bean bags that are half as heavy, how many can she make?

4. A restaurant's small salt shakers contain 0.6 ounces of salt. Its large shakers hold twice as much. The shakers are filled from a container that has 18.6 ounces of salt. If 8 large shakers are filled, how many small shakers can be filled with the remaining salt?

1. Estimate, and then divide.

I can think of multiplying both the dividend (89.6) and the divisor (0.8) by 10 to get 896 ÷ 8.

a. $89.6 \div 0.8 \approx \mathbf{880 \div 8 = 110}$

$= \frac{89.6}{0.8}$

I can multiply this fraction by 1, or $\frac{10}{10}$, to get a denominator that is a whole number.

$= \frac{89.6 \times 10}{0.8 \times 10}$

$= \frac{896}{8}$

$= 112$

$$\begin{array}{r} 112 \\ 8\,\overline{)\,896} \\ -\underline{8} \\ 09 \\ -\underline{8} \\ 16 \\ -\underline{16} \\ 0 \end{array}$$

I use the long division algorithm to solve 896 divided by 8. The answer is 112, which is very close to my estimated answer of 110.

I'll imagine multiplying both the dividend and the divisor by 100 to get 524 ÷ 4.

b. $5.24 \div 0.04 \approx \mathbf{400 \div 4 = 100}$

$= \frac{5.24}{0.04}$

I can multiply this fraction by 1, or $\frac{100}{100}$, to get a denominator that is a whole number.

$= \frac{5.24 \times 100}{0.04 \times 100}$

$= \frac{524}{4}$

$= 131$

524 divided by 4 is equal to 131.

$$\begin{array}{r} 131 \\ 4\,\overline{)\,524} \\ -\underline{4} \\ 12 \\ -\underline{12} \\ 04 \\ -\underline{4} \\ 0 \end{array}$$

2. Solve using the standard algorithm. Use the thought bubble to show your thinking as you rename the divisor as a whole number.

$2.64 \div 0.06 = \mathbf{44}$

I write a note explaining how I can rewrite the division expression from $2.64 \div 0.06$ to $264 \div 6$. Both expressions are equivalent.

I multiplied 2.64 and 0.06 by 100 to get an equivalent division expression with whole numbers.

$$\mathbf{2.64 \div 0.06 = \frac{264}{6}}$$

$$\begin{array}{r} 44 \\ 6\overline{)264} \\ -\underline{24} \\ 24 \\ -\underline{24} \\ 0 \end{array}$$

I solve by using the long division algorithm, $264 \div 6 = 44$.

Name ______________________________ Date ________________

1. Estimate and then divide. An example has been done for you.

$78.4 \div 0.7 \approx 770 \div 7 = 110$

$= \frac{78.4}{0.7}$

$= \frac{78.4 \times 10}{0.7 \times 10}$

$= \frac{784}{7}$

$= 112$

$$\begin{array}{r} 112 \\ 7\overline{)784} \\ -7 \\ 8 \\ -7 \\ 14 \\ -14 \\ 0 \end{array}$$

a. $61.6 \div 0.8 \approx$

b. $5.74 \div 0.7 \approx$

2. Estimate and then divide. An example has been done for you.

$7.32 \div 0.06 \approx 720 \div 6 = 120$

$= \frac{7.32}{0.06}$

$= \frac{7.32 \times 100}{0.06 \times 100}$

$= \frac{732}{6}$

$= 122$

$$\begin{array}{r} 122 \\ 6\overline{)732} \\ -6 \\ 13 \\ -12 \\ 12 \\ -12 \\ 0 \end{array}$$

a. $4.74 \div 0.06 \approx$

b. $19.44 \div 0.54 \approx$

3. Solve using the standard algorithm. Use the thought bubble to show your thinking as you rename the divisor as a whole number.

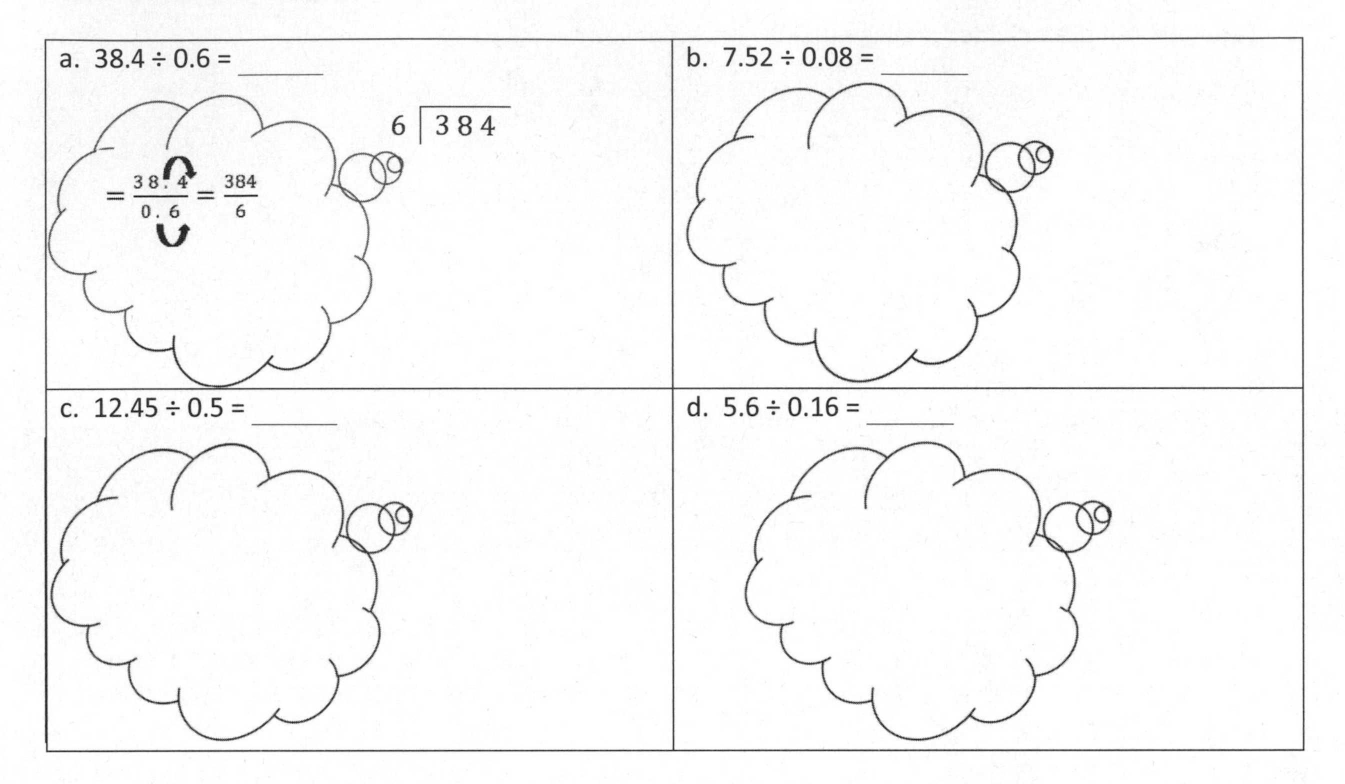

4. Lucia is making a 21.6 centimeter beaded string to hang in the window. She decides to put a green bead every 0.4 centimeters and a purple bead every 0.6 centimeters. How many green beads and how many purple beads will she need?

5. A group of 14 friends collects 0.7 pound of blueberries and decides to make blueberry muffins. They put 0.05 pound of berries in each muffin. How many muffins can they make if they use all the blueberries they collected?

1. Circle the expression equivalent to *the sum of* 5 *and* 2 *divided by* $\frac{1}{5}$.

$\frac{5+2}{5}$ $5+\left(2\div\frac{1}{5}\right)$ $\frac{1}{5}\div(5+2)$ $(5+2)\div\frac{1}{5}$ (circled)

This expression represents the sum of 5 and 2 divided by 5.

This expression represents the sum of 5 and the quotient of 2 divided by $\frac{1}{5}$.

This expression represents $\frac{1}{5}$ divided by the sum of 5 and 2.

This expression is equivalent to the sum of 5 and 2 divided by $\frac{1}{5}$.

2. Fill in the chart by writing an equivalent numerical expression.

I can find "half" by dividing by 2 or by multiplying by $\frac{1}{2}$.

The *difference* between two numbers means I need to use subtraction to solve.

This is one possible way to write the numerical expression.

a.	Half as much as the difference between $1\frac{1}{4}$ and $\frac{5}{8}$	$\left(1\frac{1}{4}-\frac{5}{8}\right)\div 2$
b.	Add 3.9 and $\frac{5}{7}$, and then triple the sum.	$\left(3.9+\frac{5}{7}\right)\times 3$

Add two numbers means I need to use addition.

I can triple a number by adding it 3 times or by multiplying by 3.

3. Fill in the chart by writing an equivalent expression in word form.

I see the subtraction sign, so I use the phrase, "*difference between* $\frac{3}{5}$ *and* ________."

I see the multiplication sign, so I use the phrase "*product of* $\frac{1}{4}$ *and* 2 *tenths*."

a.	***The difference between*** $\frac{3}{5}$ ***and the product of*** $\frac{1}{4}$ ***and*** **2** ***tenths***	$\frac{3}{5} - \left(\frac{1}{4} \times 0.2\right)$
b.	$\frac{3}{2}$ ***times the sum of*** **2.75** ***and*** $\frac{1}{8}$	$\left(2.75 + \frac{1}{8}\right) \times \frac{3}{2}$

I see the addition sign, so I use the phrase "*sum of* 2.75 *and* $\frac{1}{8}$."

I see the multiplication symbol, so I say, "$\frac{3}{2}$ *times*."

Evaluate means to "find the value of."

4. Evaluate the following the expression.

I see two multiplication signs in this expression, so I can solve for it from left to right. But since multiplication is associative, I can solve $\frac{4}{9} \times \frac{9}{4}$ first because I can see that the product is 1.

$\frac{1}{2} \times \frac{4}{9} \times \frac{9}{4}$

I put a parenthesis around $\frac{4}{9} \times \frac{9}{4}$ to show that I solve it first.

$= \frac{1}{2} \times \left(\frac{4}{9} \times \frac{9}{4}\right)$

$\frac{4}{9} \times \frac{9}{4}$ is equal to $\frac{36}{36}$, or 1.

$= \frac{1}{2} \times 1$

$= \frac{1}{2}$

$\frac{1}{2}$ of 1 is $\frac{1}{2}$.

Name ______________________________ Date ________________

1. Circle the expression equivalent to *the difference between 7 and 4, divided by a fifth.*

$7 + (4 \div \frac{1}{5})$ $\frac{7-4}{5}$ $(7-4) \div \frac{1}{5}$ $\frac{1}{5} \div (7-4)$

2. Circle the expression(s) equivalent to *42 divided by the sum of* $\frac{2}{3}$ *and* $\frac{3}{4}$.

$(\frac{2}{3} + \frac{3}{4}) \div 42$ $(42 \div \frac{2}{3}) + \frac{3}{4}$ $42 \div (\frac{2}{3} + \frac{3}{4})$ $\frac{42}{\frac{2}{3} + \frac{3}{4}}$

3. Fill in the chart by writing the equivalent numerical expression or expression in word form.

	Expression in word form	Numerical expression
a.	A fourth as much as the sum of $3\frac{1}{8}$ and 4.5	
b.		$(3\frac{1}{8} + 4.5) \div 5$
c.	Multiply $\frac{3}{5}$ by 5.8; then halve the product	
d.		$\frac{1}{6} \times (4.8 - \frac{1}{2})$
c.		$8 - (\frac{1}{2} \div 9)$

4. Compare the expressions in 3(a) and 3(b). Without evaluating, identify the expression that is greater. Explain how you know.

5. Evaluate the following expressions.

a. $(11 - 6) \div \frac{1}{6}$

b. $\frac{9}{5} \times (4 \times \frac{1}{6})$

c. $\frac{1}{10} \div (5 \div \frac{1}{2})$

d. $\frac{3}{4} \times \frac{2}{5} \times \frac{4}{3}$

e. 50 divided by the difference between $\frac{3}{4}$ and $\frac{5}{8}$

6. Lee is sending out 32 birthday party invitations. She gives 5 invitations to her mom to give to family members. Lee mails a third of the rest, and then she takes a break to walk her dog.

a. Write a numerical expression to describe how many invitations Lee has already mailed.

b. Which expression matches how many invitations still need to be sent out?

$32 - 5 - \frac{1}{3}(32 - 5)$ $\frac{2}{3} \times 32 - 5$ $(32 - 5) \div \frac{1}{3}$ $\frac{1}{3} \times (32 - 5)$

I can represent this story with the expression $\frac{1}{4} \div 3$.

1. Mrs. Brady has $\frac{1}{4}$ liter of juice. She distributes it equally to 3 students in her tutoring group.

    a. How many liters of juice does each student get?

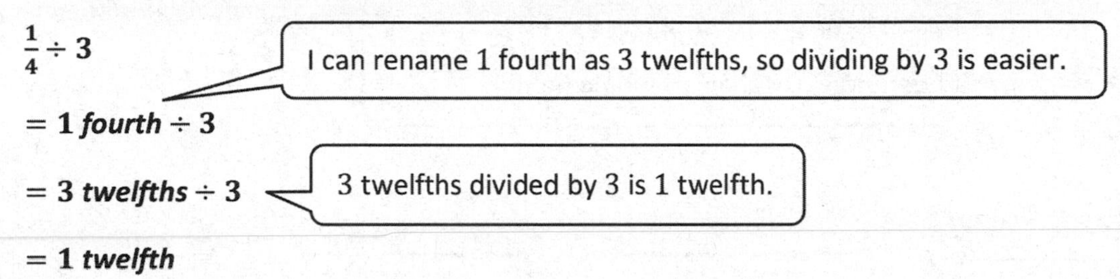

Each student gets $\frac{1}{12}$ ***liter of juice.***

    b. How many more liters of juice will Mrs. Brady need if she wants to give each of the 36 students in her class the same amount of juice found in Part (a)?

$36 \times \frac{1}{12}$ *liter*

I can multiply to find how much juice she'll need to serve 36 students.

$= \frac{36 \times 1}{12}$ *liters*

$= \frac{36}{12}$ *liters*

Mrs. Brady will need 3 liters of juice for 36 students.

$= 3$ *liters*

3 *liters* $- \frac{1}{4}$ *liter* $= 2\frac{3}{4}$ *liters*

I subtract to find out how much more juice she'll need.

Mrs. Brady will need an additional $2\frac{3}{4}$ ***liters of juice.***

2. Austin buys $16.20 worth of grapefruit. Each grapefruit costs $0.60.

   a. How many grapefruits does Austin buy?

   $\$16.20 \div \$0.60$

   $= \frac{16.2}{0.6} \times \frac{10}{10}$

   $= \frac{162}{6}$

   $= 27$

   To find how many grapefruits Austin buys, I use the total cost divided by the cost of each grapefruit.

   I multiply the fraction by 1, or $\frac{10}{10}$, to get a denominator that is a whole number.

$$
\begin{array}{r}
27 \\
6\,\overline{)\,162} \\
-\,12 \\
\hline
42 \\
-\,42 \\
\hline
0
\end{array}
$$

   I use the long division algorithm to solve 162 divided by 6. The answer is 27.

   Austin buys 27 grapefruits.

   b. At the same store, Mandy spends one third as much money on grapefruit as Austin. How many grapefruits does she buy?

   Since Mandy spent $\frac{1}{3}$ as much money on grapefruit as Austin, that means she's buying $\frac{1}{3}$ the number of grapefruit.

   $27 \div 3 = 9$

   Mandy buys 9 grapefruits.

   To find one third of a number, I can multiply by $\frac{1}{3}$ or divide by 3.

Name ______________________________ Date ______________

1. Chase volunteers at an animal shelter after school, feeding and playing with the cats.

   a. If he can make 5 servings of cat food from a third of a kilogram of food, how much does one serving weigh?

   b. If Chase wants to give this same serving size to each of 20 cats, how many kilograms of food will he need?

2. Anouk has 4.75 pounds of meat. She uses a quarter pound of meat to make one hamburger.

   a. How many hamburgers can Anouk make with the meat she has?

   b. Sometimes Anouk makes sliders. Each slider is half as much meat as is used for a regular hamburger. How many sliders could Anouk make with the 4.75 pounds?

3. Ms. Geronimo has a $10 gift certificate to her local bakery.

   a. If she buys a slice of pie for $2.20 and uses the rest of the gift certificate to buy chocolate macaroons that cost $0.60 each, how many macaroons can Ms. Geronimo buy?

   b. If she changes her mind and instead buys a loaf of bread for $4.60 and uses the rest to buy cookies that cost $1\frac{1}{2}$ times as much as the macaroons, how many cookies can she buy?

4. Create a story context for the following expressions.

   a. $(5\frac{1}{4} - 2\frac{1}{8}) \div 4$

   b. $4 \times (\frac{4.8}{0.8})$

5. Create a story context for the following tape diagram.

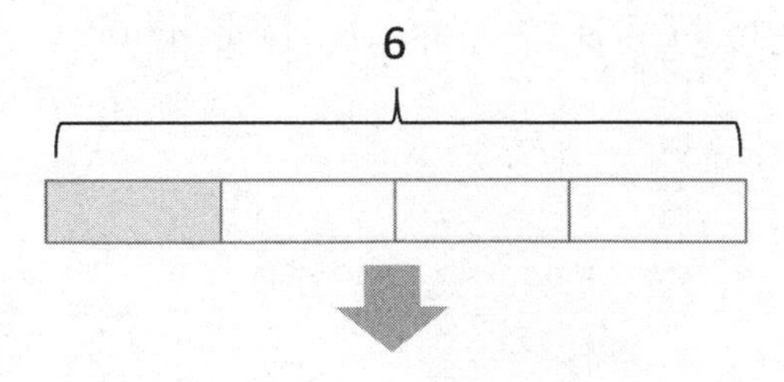

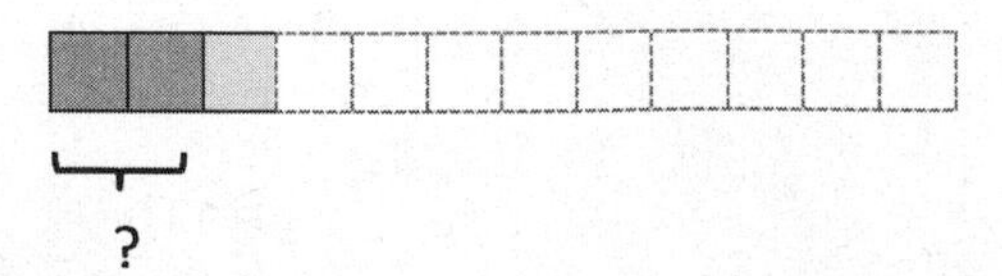

Credits

Great Minds® has made every effort to obtain permission for the reprinting of all copyrighted material. If any owner of copyrighted material is not acknowledged herein, please contact Great Minds for proper acknowledgment in all future editions and reprints of this module.